The Theory and Practice of Brewing

Michael Combrune

Alpha Editions

This edition published in 2023

ISBN : 9789357948159

Design and Setting By
Alpha Editions
www.alphaedis.com
Email - info@alphaedis.com

As per information held with us this book is in Public Domain.
This book is a reproduction of an important historical work. Alpha Editions uses the best technology to reproduce historical work in the same manner it was first published to preserve its original nature. Any marks or number seen are left intentionally to preserve its true form.

THE PREFACE.

THE difference that appears in the several processes of brewing, though executed with the same materials, by the same persons, and to the same intent, is generally acknowledged. The uneasiness this must occasion to those who are charged with the directive part of the business, cannot be small: and the more desirous they are of well executing the duty incumbent on them, the greater is their disappointment, when frustrated in their hopes. To remove this uncertainty, no method seems preferable to that of experiments, as it is by this means alone, any art whatever can be established upon a solid foundation: but these require caution, perseverance, and expence; they must be multiplied and varied both for the same and for different purposes. The operations of nature elude superficial enquiries, where we have few or no principles for our guides, many experiments are made, which tend only to confound or deceive. Effects seen, without a sufficient knowledge of their causes, often are neglected, or viewed in an improper light, seldom faithfully reported, and, for want of distinguishing the several circumstances that attend them, many times become the support of old prejudices, or the foundation of new ones.

Whoever is attentive to the practical part of brewing, will soon be convinced that heat, or fire, is the principal agent therein, as this element, used in a greater or less degree, or differently applied, is the occasion of the greatest part of the variety we perceive. It is but a few years since the thermometer has been found to be an instrument sufficiently accurate for any purposes where the measure of heat is required. And, as it is the only one with which we are enabled to examine the processes of brewing, and to account for the difference in the effects, a theory of the art, founded on practice, must be of later date than the discovery of the instrument that guides us to the principles.

So long since as the year 1741, I began this research, and never neglected any opportunity to consult the artists of the trade, or to try such experiments as I conceived might be conducive to the purpose. It is needless, perhaps shameful, to mention their number, or to speak of the many disappointments I met with in this pursuit. Error admits of numberless combinations. Truth alone is simple, and confirmed by continuity. At last, flattering myself with having collected the true theory, assisted and encouraged by men of abilities, I thought it fit the public should judge whether I had succeeded in my enxideavours; and in 1758 the Essay on Brewing was submitted to them, either for their approbation, or that the errors therein might be pointed out. I have had no reason to repent of my temerity, though perhaps the novelty, more than the merit of this performance, engaged the attention, I may add the favor and advice of some good judges. They have allowed my principles to be at least plausible, and their agreement with practice has since repeatedly convinced me they were not far from truth.

The Essay just mentioned, revised and corrected, naturally forms the first part or theory of the present treatise. The second part is entirely practical. After giving a short idea of the whole process, I resume its different branches in as many chapters, and endeavour in such manner to guide the practitioner, that he may, in every part, at all times, and under a variety of circumstances, know what he is to do, and seldom, if ever, to be disappointed in his object.

From the investigation of so extensive a business, some benefit, it is hoped, must accrue to the public; from the process of brewing being carried on in a just and uniform manner, our malt liquors, probably, will in time better deserve the name of wine.

Boerhaave, Shaw, Macquer, and most of the great masters in chymistry are far from limiting that name to the liquors produced from the juice of the grape: they extend it to all fermented vegetable juices, which, on distillation, yield an ardent spirit, and look on the strength and faculty wine has to cherish nature, and preserve itself, to be in proportion to the quantity it possesses of this liquid, generally termed spirit of wine. This, when thoroughly pure and dephlegmated, is one and the same, whatever different vegetable it is produced from. Barley wines possess the same spiritous principle, which is the preservative part of the most valuable foreign wines, with a power of being brewed superior or inferior to them in quality, and the other constituent parts of beer, beside this ardent spirit, will not, I believe, be esteemed less wholesome than those which make up the whole of grape wine.

The reasons why Great Britain hath not hitherto furnished foreign nations with this part of her product, but more especially her seamen, are obvious. Our mariners, when at home, do not dislike beer, either as to their palates, or its effects on their constitution; but when abroad, spiritous liquors, or new wines, often the product of an enemy's country, are substituted in lieu thereof. The disuse of beers, on these occasions, has been owing to the uncertainty of the principles on which they were brewed; the maintaining them sound in long voyages and in hot climates, could not sufficiently be depended upon; and it has been supposed they could not be procured at so easy a rate as wines, brandies, or rums, purchased abroad. The first of these objections, the author hopes, by this work, to remove; and, were all the duties to be allowed on what would be brewed for this purpose, our seamen might be furnished with beer stronger than Spanish wine, and at a less expence, the mean price of malt and hops being taken for seven years. It is true that, in times of peace, the seamen in his Majesty's service are not very numerous, but the number of those then employed by merchants is considerable. I should not have presumed to mention this, but on account of the encouragement given to the exportation of corn, and to many manufactures of British growth or British labor. It is computed that, in England and Wales, are brewed three millions five hundred thousand quarters of malt yearly, for which purpose upwards of one hundred and fifty thousand weight of hops are used. The improvement of the brewery might become a means of increasing the consumption of the growth of our country, viz. of barley, to more than one hundred thousand quarters, and of hops to between fourteen and fifteen thousand weight annually.

Whether this be an object deserving the attention of the legislative power, or of the landed interest, and what might be the proper means to put it successfully in practice, are considerations which do not belong to this place; it being sufficient here to point out, how universally beneficial it is to establish the art of brewing on true and invariable principles.

This being the first attempt, that has been made, to reduce this art to rules and principles, the Author hopes he has a just claim to the indulgence of the public, for any errors he unwillingly may have adopted; far from believing that there is no room left for future improvements, he recommends it to those, who, blessed with superior talents and more leisure than himself, may be inclined to try their skill in the same field, to watch closely the steps of NATURE; after the strictest enquiry made, it will be found, the success of brewing beers and ales wholly depends on a true imitation of the wines she forms.

This second edition, it may be observed, in many respects, differs considerably from the first. I have endeavoured to convert to use every advice, every opinion I received, and having put these to the test of farther practice, flatter myself it will be found improved.

AN EXPLANATION OF THE TECHNICAL TERMS.

THE intent of every brewer, when he forms his drink, is to extract the fermentable parts of the malt, in the most perfect manner; to add hops, in such proportion as experience teaches him will preserve and ameliorate the beer; and to employ just so much yeast as is sufficient to obtain a complete fermentation.

Perhaps it may be said, these particulars are already sufficiently understood, and that it would be a much more useful work to publish remedies for the imperfections, or diseases, beer is naturally or accidentally subject to, and which at present are deemed incurable. But if the designs just now mentioned be executed according to the rules of chymistry, such imperfections and such diseases not existing, the remedies will not be wanted; for beer brewed upon true principles, is, neither naturally nor accidentally, subject to many disorders often perceived in it. Hence it is evident, that some knowledge of chymistry is absolutely necessary to complete the brewer, as, without the informations acquired from that science, he must be unqualified to lay down rules for his practice, and to secure to himself the favor of the public; for which purpose, and to make this treatise useful to those concerned in the practical part of brewing, it has been thought adviseable to avoid, as much as possible, the technical terms of art, to prefix an explanation of those that necessarily occur, and, in as short a manner as possible, to trace the properties of fire, air, water, and earth, as far as they relate to the subject.

ACIDS are all those things which taste sour, as vinegar, juice of lemons, spirit of nitre, spirit of salt, the oil and spirit of vitriol, &c. and are put in a violent agitation, by being mixed with certain earths, or the ashes of vegetables. An acid enters, more or less, into the composition of all plants, and is produced by, or rather is the last effect of, fermentation. Mixed in a due proportion with an alkali, it constitutes a neutral salt, that is, a salt wherein neither the acid nor alkali prevail. Acids are frequently termed acid salts, though generally they appear under a fluid form.

ALKALIES, or alkaline salts, are of a nature directly contrary to the acids, and generally manifest themselves by effervescing therewith: they have an urinous taste, and are produced from the ashes of vegetables, and by several other means. They, as well as testaceous and calcarious substances, are frequently made use of by coopers, to absorb the acid parts of stale beer, by them called *softning*.

AIR is a thin elastic fluid, surrounding the globe of the earth; it is absolutely necessary to the preservation both of animal and vegetable life, and for the exciting and carrying on fermentation.

ALCOHOL is the pure spirit of wine, generally supposed to be without the least particle of water or phlegm.

ANIMALS are organized bodies, endued with sensation and life. Minerals are said to grow and increase, plants to grow and live, but animals only to have sensation.—Animal substances cannot ferment so as to produce by themselves a vinous liquor; but there may be cases wherein some of their parts rather help than retard the act of fermentation.[1]

ATMOSPHERE is that vast collection of air, with which the earth is surrounded to a considerable height.

ATTRACTION is an indefinite term, applicable to all actions whereby bodies tend towards one another, whether by virtue of their weight, magnetism, electricity, or any other power. It is not, therefore, the cause determining some bodies to approach one another, that is expressed by the word *attraction*, but the effect itself. The space, through which this power extends, is called the *sphere of attraction*.

BLACKING is a technical term used by coopers, to denote sugar that is calcined, until it obtains the colour that occasions the name.

BREWING is the operation of preparing beers and ales from malt.

BOILING may thus be accounted for. The minute particles of fuel being by fire detached from each other, and becoming themselves fire, pass through the pores of the vessel, and mix with the fluid. These, being perpetually in an active state, communicate their motion to the water: hence arises, at first, a small intestine motion, and from a continued action in the first cause, the effect is increased, and the motion of the liquor continually accelerated; by degrees, it becomes sensibly agitated, but the particles of the fire, acting chiefly on the particles that compose the lowest surface of the water, give them an impulse upwards, by rendering them specifically lighter, so as to determine them to ascend, according to the laws of equilibrium. Hence there is a constant flux of water from the bottom to the top of the vessel, and reciprocally from the top to the bottom. This appears to be the reason why water is hot at the top sooner than at the bottom, and why an equal heat cannot be distributed through the whole. The thermometer therefore can be of little service, to determine immediately the degree of heat, especially in large vessels, on which account it is better for brewers to heat a certain quantity just to the act of boiling, and to temper it, by adding a sufficient quantity of cold water. Boiling water is incapable of receiving any increase of heat, though acted on by ever so great a fire, unless the atmosphere becomes heavier, or the vapours of the water be confined. It occasions the mercury to rise, according to Farenheit's scale, to 212 degrees.

CHARR. A body is said to be charred when, by fire, its volatile or most active parts are drove out; its coarse oils, by the same means, placed chiefly on the external parts; and so deprived of color as to be quite black.

CLEANSING is the act of removing the beer from the ton, where it was first fermented, into the casks.

CLOUDY is an epithet joined to such beers, which, from the violent heat given to the water that brewed them, are loaded with more oils than can be attenuated by fermentation, and incorporated with the water; from whence a muddy and grey oil is seen floating on the surface of the liquor, though the body is often transparent; this oil is frequently extracted in such quantity as to exceed the power of any known menstruum.

COHESION is that action by which the particles of the same body adhere together, as if they were but one.

COLD is a relative term in opposition to heat. Its greatest degree is not known, and it is supposed that the colder a body is, the less is the agitation of its internal parts.

COLOUR; a greater or less degree of heat causes different colours in most bodies, and from a due observation of the colour of malt, we may determine what degree of heat it has been impressed with.

DENSITY expresses the closeness, compactness, or near approach of the parts of a body to one another: the more a body weighs in proportion to its bulk, the greater is its density. Gold is the densest body in nature, because there is none known of the same bulk, which weighs so much.

EARTH is that fossil matter or element, whereof our globe partly consists.

EBULLITION is the boiling or bubbling of water, or any other liquor, when the fire has forced itself a passage through it. Brewers suppose water to be just beginning to boil, when they perceive a small portion of it forced from the bottom upwards in a right line, so as to disturb the surface: when the liquor is in this state, they call it *through*, or upon the point of ebullition. The vulgar notion that the water is hotter at this time than when it boils, is without any foundation.

EFFERVESCENCE is a sudden agitation, arising in certain bodies upon mixing them together; this agitation most commonly generates heat.

ELASTICITY, or springiness, is that property of bodies, by which they restore themselves to their former figure, after any pressure or distension.

EXPANSION is the swelling or increase of the bulk of bodies from heat, or any other cause.

EXTRACT consists of the parts of a body separated from the rest, by cold or hot water.

FERMENTATION is a sensible internal motion of the particles of a mixture: by the continuance of this motion, the particles are gradually removed from their former situation, and, after some visible separation, joined together again in a different order and arrangement, so as to constitute a new compound. No liquors are capable of inebriating, except those that have been fermented.

FIXED BODIES are those, which, consisting of grosser parts, cohering by a strong attraction, and by that means less susceptible of agitation, can neither be separated nor raised, without a strong heat, or perhaps not without fermentation.

FIRE is only known by its properties, of which the chief are to penetrate and dilate all solid and fluid bodies.

FREEZING POINT is the degree of cold, at which water begins to be formed into ice, which, according to Farenheit's scale, is expressed by 32.

FOXED is a technical term, used by brewers, to indicate beers in a putrid state.

GUMS are concreted vegetable juices, which transude through the bark of certain trees, and harden upon the surface; they easily dissolve in water, and by that means distinguish themselves from balsams or resins.

HERMETICALLY SEALED is a particular method of stopping the mouth of vessels, so close that the most subtil spirit cannot fly out, which is done by heating the neck of the bottles, till it is just ready to melt, and then with hot pinchers twisting it close together.

HOMOGENEOUS is an appellation given to such parts or subjects, which are similar or of the same nature and properties.

ISINGLASS is a preparation from a fish called huso, somewhat bigger than the sturgeon; a solution of which in stale beer is used, to fine or precipitate other beers: it is imported from Russia by the Dutch, and from them to us.

LIGHT consists of particles of matter inconceivably small, capable of exciting in us the sensation of colours, by being reflected from every point of the surface of luminous bodies; but, notwithstanding they are so exceeding small, Sir Isaac Newton found means to divide a single ray into seven distinct parts, viz. red, orange, yellow, green, blue, indigo, and violet.

MALT, in general, is any sort of grain, first germinated, and then dried, so as to prevent any future vegetation: that generally used, is made of barley, which experience has found to be the fittest for the purpose of brewing.

MEDIUM is that space, through which a body in motion passes: air is the medium through which the bodies near the earth move; water is the medium wherein fish live; glass affords a medium or a free passage to light.—This term is also made use of, to express the mean of two numbers, and sometimes the middle between several quantities.

MUSTS are the unfermented juices of grapes, or of any other vegetable substances.

MENSTRUUM is any fluid, which is capable of interposing its parts between those of other bodies, and in this manner either dissolves them perfectly, or extracts some part of them.

OIL is an unctuous, inflammable substance, drawn from several animal and vegetable substances.

PRECIPITATION. Isinglass dissolved becomes a glutinous and heavy body; this put into malt liquors intended to be fined, carries down, by its weight, all those swimming particles, which prevent its transparency; and this act is called fining, or precipitation.

REPULSION; "Doctor Knight defines it to be that cause which makes bodies mutually endeavour to recede from each other, with different forces at different times." In this case they are placed beyond the sphere of each other's attraction or cohesion, and mutually fly from each other.

RESINS, or balsams, are the oils of vegetables inspissated and combined with a proportion of the acid salts; as well as they mix with any spirituous liquor, as little are they soluble in water; but they become so, either by the intervention of gums or soaps, or by the attenuating virtue of fermentation.

SALTS are substances sharp and pungent, which readily dissolve in water, and from thence, by evaporation, crystallise and appear in a solid form. They easily unite together, and form different compounds. Thus salts, composed of acids and alkalies, partake of both, and are called neutral.

SETT: a grist of malt is by brewers said to be sett, when, instead of separating for extraction, it runs in clods, increases in heat, and coagulates. This accident is owing to the over quantity of fire in the water, applied to any of the extractions. The air included in the grist, which is a principal agent in resolving the malt, being thereby expelled, the mass remains inert, and its parts, adhering too closely together, are with difficulty separated. Though an immediate application of more cold water to the grist is the only remedy, yet, as the cohesion is speedy and strong, it seldom takes effect.—New malts, which have not yet lost the heat they received from the kiln, are most apt to lead the brewer into this error, and generally in the first part of the process.

SUGAR, or saccharine salts, are properly those that come from the sugar canes; many plants, fruits and grains give sweet juices reducible to the same form; they are supposed to be acids smoothed over with oils; all vegetable sweets are capable of fermenting spontaneously when crude; if boiled, they require an addition of yeast to make them perform that act. Malt, or its extracts, have all the properties of saccharine salts.

SULPHUR. Though by sulphur is commonly understood the mineral substance called brimstone, yet in chymistry it is frequently used to signify in general any oily substance, inflammable by fire, and, without some saline addition, indissoluble in water.

SOAP OR SAPONACEOUS JUICES. Common soap is made of oil mixed with alkaline salts: this mixture causes a froth on being agitated in water. The oils of vegetables are, in some degree, mixed with their salts; and according to the nature of these salts, appear either resinous or saponaceous, that is, soluble or indissoluble in water.— Sugar is a kind of soap, rendering oil miscible with water; and therefore all bodies, from which saccharine salts are extracted, may be termed saponaceous.

VEGETABLE is a term applied to plants, considered as capable of growth, having vessels and parts for this purpose, but generally supposed to be without sensation.

VINEGAR is an acid penetrating liquor, prepared from wine, beer, cyder, or a must, which has been fermented as far as it was capable.

VITRIOL is, in general, a metalline substance combined with the strongest acid salt known. This acid, being separated from the metal, differs in nothing from that which is extracted from alum or brimstone. It is improperly called spirit of vitriol, when diluted with water, and, with as little propriety, oil, when free from it.

VOLATILE BODIES are those, which, either from their smallness or their form, do not cohere very strongly together, and being most susceptible of those agitations, which keep liquors in a fluid state, are most easily separated and rarified into vapour, with a gentle heat, and on the contrary condensed and brought down with cold.

WINE is a brisk, agreeable, spirituous, fluid cordial, formed from fermented vegetable bodies. In this sense beers and ales may be called, and really are, barley wines.

WORTS are the unfermented extracts of malt.

YEAST is both the flowers and lees of a fermented wort, the former of these being elastic air enveloped in a subject less strong and less consistent than the latter.

PRINCIPLES OF THE THEORY OF BREWING.

SECTION I.

OF FIRE.

THOUGH fire is the chief cause and principle of almost every change in bodies, and though persons untaught in chymistry imagine they understand its nature, yet, certain it is, few subjects are so incomprehensible, or elude so much our nicest research. The senses are very inadequate judges of it; the eye may be deceived, and suppose no fire in a bar of iron, because it does not appear red, though at the same time it may contain enough to generate pain: the touch is equally unfaithful, for a body, containing numberless particles of heat, will to us feel cold, if it is much more so than ourselves.

The great and fundamental difference among philosophers, in respect to the nature of fire, is, whether it be originally such, formed by the Creator himself, at the beginning of things; or whether it be mechanically producible in bodies, by inducing some alteration in the particles thereof. It is certain that heat may be generated in a body, by attrition; but whether it existed there before, or was caused immediately by the motion, is a matter of no great import to the art of brewing; for the effects, with which we are alone concerned, are the same.

Fire expands all bodies, both solid and fluid. If an iron rod just capable of passing through a ring of the same metal, is heated red-hot, it will be increased in length, and so much swelled as not to be able to pass through the ring, as before:[2] if a fluid is put into a bellied glass, with a long slender neck, and properly marked, the fluid, by being heated, will manifestly rise to a considerable height.

The expansion of fluids, by heat, is different in different fluids; with some exceptions, it may be said to be in proportion to their density. Pure rain water, gradually heated to ebullition, is expanded one 26th part of its bulk,[3] so that 27 gallons of boiling water, will, when cold, measure no more than 26, and 27 gallons of boiling wort will not yield so much, because worts contain many oily particles, which, though less dense than water, have the property of being more expansible: hence we see the reason why a copper, containing a given number of barrels of wort, when cold, is not capable to hold the same of beer, when boiling.

Bodies are weakened or loosened in their texture by fire: the hardest, by an increased degree of heat, will liquify and run; and vegetables are resolved and separated by it into their constituent parts. It must be owned vegetables seem at first, on being exposed to the fire, to become rigid or stiff; but this is owing to the evaporation of the aqueous particles, which prevented a closer adhesion of the solid matter. It is only in this manner fire strengthens some bodies which before were weak.

That the texture of bodies should be loosened by fire, seems a consequence of expansion; for a body cannot be expanded but by its particles receding farther from one another; and if these be not able to regain the situation they had when cold, the body will remain looser in its texture than before it suffered the action of fire. This is the case of barley when malted.

Fire may be conveyed through most bodies, as air, water, ashes, sand, &c. The effect seems to be different according to the different conveyances. A difference appears between boiling and roasting, yet they answer the same purpose, that of preserving the subject; and this, in proportion to the degree of heat it has suffered. A similar variety appears, even to our taste, from the different conveyance of fire to malt: for acids having a great tendency to unite with water, if this element does not naturally contain any itself, is the reason why a great heat is conveyed through water, and applied to extract the virtues of pale malt; the water gaining from the grain some of these salts, or possessing them itself, the effect of this great aqueous heat is not to imprint on the palate a nauseous burnt taste, as is the case of great heats, when conveyed through air to the same grain. The salts the water has obtained, or perhaps had, being sheathed by the oils it draws from the malt, rather become saccharine, which cannot be the case when oils are acted upon by a strong heat, entirely void of any such property; but malt, the more it is dried, the longer is it capable of maintaining itself in a sound state, and the liquor brewed with it will, in proportion to its dryness, keep the longer sound, the hotter the water is, applied to malt, provided its heat doth not exceed the highest extracted degree, the more durable and sound will the extract be.

The last consideration of fire or heat, relative to brewing, is the knowledge of its different degrees, and how to regulate them. Till of late, chymists and all others, were much to seek in this respect; they distinguished more or less fire in a very vague and indeterminate manner, as the first, second, third, and fourth degree of heat, meaning no precise heat, or heat measured by any standard; but, by the invention of the thermometer, we are enabled to regulate our fires with the utmost precision. Thermometers are formed on different scales; and therefore, when any degree of heat is mentioned, in order to avoid confusion, the scale made use of should be indicated. I have constantly employed Fahrenheit's, as it is the most perfect, and the most generally received. According to this instrument,[4] by the author of it, an artificial cold was made so as the mercury stood at 72 divisions below the first frost. The gentlemen of the French Academy, in the winter of the year

1736, observed, at Torneao, Latitude 65° 51´, the natural cold to be 33 degrees below 0: these are proofs there are colds much more intense than the first frost, or 32 degrees, where water first begins to harden into ice; from 32 to 90 degrees are the limits of vegetation, according to the different plants that receive those or the intermediate heats. The 40th degree is marked by Boerhaave as the first fermentable heat, and the 80th as the last: 47 degrees I have found to be generally the medium heat of London, throughout the year, in the shade; 98 degrees is said to be that of our bodies when in health, as from 105 to 112 are its degrees when in a fever. Hay stacked with too much moisture, when turned quite black, in the heart of the rick, indicated a heat of 165 degrees. At 175 the purest and highest-rectified spirits of wine boil, and at this degree I have found well-grown malts to charr, at 212 degrees water boils, at 600 quicksilver and oil of vitriol. Gold, silver, iron, and most other metals in fusion exceed this heat; greater still than any known is the fire in the focus of the burning lens of Tschirnhausen, or of the concave mirror made by Villette; they are said to volatilise metals and vitrify bricks. Thus far experiments have reached; but how much more, or how much less, the power of this element extends, will probably be forever hid from mankind.

SECTION II.

OF AIR.

NONE of the operations, either of nature or art, can be carried on without the action or assistance of air. It is a principal agent in fermentation; and therefore brewers ought to be well acquainted with its principal properties and powers.

By air we mean a fluid, scarcely perceptible to our senses, and discovering itself only by the resistance it makes to bodies. We find it every where incumbent on the surface of the globe, rising to a considerable height, and commonly known by the name of atmosphere. The weight of air is to that of water as 1 to 850, and its gravitating force equal to that of a column of water of 33 feet high; so that an area of one foot square receives, from air, a pressure equal to 2080 pounds weight.

Elasticity is a property belonging only to this element, and this quality varies in proportion to the compressing weights. We scarcely find this element, (any more than the others) in a pure state; one thousandth part of common air, says Boerhaave, consists of aqueous, spiritous, oily, saline, and other particles scattered through it.—These are not, or but little, compressible, and in general prevent fermentation: consequently, where the air is purest, fermentation is best carried on. The same author suspects, that the ultimate particles of air cohere together, so as not easily to insinuate themselves into the smallest pores, either of solids or fluids. Hence, those acquainted with brewing, easily account, why very hot water, which forces strong and pinguious particles from malt, forms at the same time extracts unfavourable for fermentation, as oils are an obstruction to the free entrance of air; and, from an analogous reason, extracts which are much less impressed with fire, in them fermentation is so much accelerated, that the whole soon becomes sour.

Air, like other bodies, is expanded and rarified by heat, and exerts its elasticity in proportion to the number of degrees of fire it has received; the hotter therefore the season is, the more active and violent will the fermentation be.

Air abounds with water, and is perpetually penetrating and insinuating itself into every thing capable of receiving it. Its weight, or gravitating force, must necessarily produce numberless effects. The water contained in the air is rendered more active by its motion; hence the saline, gummous, and saponaceous particles it meets with are loosened in their texture, and, in some degree, dissolved. As principles similar to these are the chief constituent parts of malt, the reason is obvious why such, which are old, or have lain a proper time exposed to the influence of the air, dissolve more readily, or, in other words, yield a more copious extract than others.

All bodies in a passive state, remaining a sufficient time in the same place, become of the same degree of heat with the air itself. On this account the water, lying in the backs used by brewers, is nearly of the same degree of heat as the thermometer shews the open air in the shade to be. When this instrument indicates a cold below the freezing point, or 32 degrees, if the water does not then become ice, the reason is, because it has not been exposed long enough to be thoroughly affected by such a cold. For water does not immediately assume the same degree of temperature with the air, principally on account of its density, also from its being pumped out of deep and hot wells, from its being kept in motion, and from many other incidents. Under these circumstances, no great error can arise to estimate its heat equal to 35 degrees.

Air is not easily expelled from bodies, either solid or fluid. Water requires two hours boiling to be discharged of the greatest part of its air. That it may be thus expelled by heat appears from hence; water, if boiled the space abovementioned, instead of having any air bubbles when it is froze, as ice commonly has, becomes a solid mass like crystal.

Worts or musts, as they contain great quantities of salts and oils, require a greater degree of heat to make them boil: consequently more air is expelled from boiling worts, than from boiling water in the same time; and as air doth not instantaneously re-enter those bodies,[5] when cold, they would never ferment of themselves. Were it not for the substitute of yeast, to supply the deficiency of air lost by boiling, they would fox or putrify, for want of that internal elastic air, which is absolutely necessary to fermentation.

As air joined to water contributes so powerfully to render that fluid more active, that water which has endured fire the least time, provided it be hot enough, will make the strongest extracts.

Though there is air in every fluid, it differs in quantity in different fluids; so that no rule can be laid down for the quantity of air, which worts should contain.—Probably the quantity, sufficient to saturate one sort, will not be an adequate proportion for another.

Air in this manner encompasses, is in contact with, confines, and compresses all bodies. It insinuates itself into their penetrable passages, exerts all its power either on solids, or fluids, and finding in bodies some elements to which it has a tendency, unites with them. By its weight and perpetual motion, it strongly agitates those parts of the bodies in which it is contained, rubs, and intermixes them intimately together. By disuniting some, and joining others, it produces very singular effects, not easily accomplished by any other means.—That this element has such surprising powers, is evident from the following experiment. "Fermentable parts duly prepared and disposed in the vacuum of Mr. Boyle's air-pump will not ferment, though acted upon by a proper heat; but, discharging their air, remain unchanged."

SECTION III.

OF WATER.

AS water is perpetually an object of our senses, and made use of for most of the purposes of life, it might be imagined the nature of this element was perfectly understood: but they who have enquired into it with the greatest care, find it very difficult to form a just idea of it. One reason of this difficulty is, water is not easily separated from other bodies, or other bodies from water. Hartshorn, after having been long dried, resists a file more than iron; yet, on distillation, yields much water. I have already observed, that air is intimately mixed with, and possibly never entirely separated from it, but in a *vacuum*; how is it possible then ever to obtain water perfectly pure?

In its most perfect state, we understand it to be a liquor very fluid, inodorous, insipid, pellucid, and colourless, which, in a certain degree of cold, freezes into a brittle, hard, glassy ice.

Lightness is reckoned a perfection in water, that which weighs less being in general the purest. Hence the great difficulty of determining the standard weight it should have. Fountain, river, or well waters, by their admixture with saline, earthy, sulphureous, and vitriolic substances, are rendered much heavier than in their natural state; on the other hand, an increase of heat, or an addition of air, by varying the expansion, diminishes the weight of water. A pint of rain-water, supposed to be the purest, is said to weigh 15 ounces, 1 drachm, and 50 grains, but, for the reasons just now mentioned, this must differ in proportion as the seasons of the year do from each other.

Another property of water, which it has in common with other liquors, is its fluidity, which is so great, that a very small degree of heat, above the freezing point, makes it evaporate. Experiments to ascertain the proportion steamed away of the quantity of water used in brewing, is an object worthy of the artist's curiosity; but the purer the water is, the more readily it evaporates. Sea-water, which is supposed to contain one fortieth part of salt, more forcibly resists the power of fire, and wastes much less, than that which is pure.

The ultimate particles of this element, Boerhaave believed to be much less than those of air, as water passes through the pores and interstices of wood, which never transmit the least elastic air; nor is there, says he, any known fluid, (fire excepted, which forces itself through every subject) whose parts are more penetrating than those of water. Yet as water is not an universal dissolver, there are vessels which will contain it, though they will let pass even the thick syrup of sugar, for sugar makes its way by dissolving the tenacious and oily substance of the wood, which water cannot do.

Water, when fully saturated by fire, is said to boil, and by the impulse of that element, comes under a strong ebullition. Just before this violent agitation takes place, I have already observed, it occupies one seventy-sixth more space than when cold: so the brewer who would be exact, when he intends to reduce his liquor to a certain degree of heat, must allow for this expansion, abating therefrom the quantity of steam exhaled.

As water, by boiling, may be said to be filled or saturated with fire, so may it be with any other substance capable of being dissolved therein; but, though it will dissolve only a given quantity of any particular substance, it may, at the same time, take in a certain proportion of some other. Four ounces of pure rain water will melt but one ounce of common salt, and after taking this as the utmost of its quantity, it will still receive two scruples of another kind of salt, viz. nitre. In like manner the strongest extract of malt is capable of receiving the properties belonging to hops: but in a limited proportion. This appears from the thin bitter pelicle, that often swims on the surface of the first wort of brown beers, which commonly are overcharged with hops, by putting the whole quantity of them at first therein; the wort not being capable of suspending all that the heat dissolves, it no sooner cools but these parts rise on the top. This may serve as a hint to prevent this error, by suffering the first wort to have no more hops boiled therein than it can sustain: but as this incident must vary, in proportion to the heat of the extracts and quantity of water used, some few experiments are necessary to indicate the due proportion for the several sorts of drink. This however should always be extended to the utmost, for the first wort, which, from its nature and constituent parts, stands most in need of the preservative quality the hops impart.

Water acts very differently, as a menstruum, according to the quantity of fire it contains: consequently its heat is a point of the utmost importance with regard to brewing, and should be properly varied according to the dryness and nature of the malt, according as it is applied either in the first or last mashes, and in proportion also to the time the beer is intended to be kept. These ends, we hope to shew, are to be obtained to a degree of numerical certitude.

Nutrition cannot be carried on without water, though likely water itself is not the matter of nourishment, but only the vehicle.

Water is as necessary to fermentation as heat or air. The farmer, who stacks his hay or corn before it is sufficiently dried, soon experiences the terrible effects of too much moisture, or water, residing therein: all vegetables therefore intended to be long kept, ought to be well dried. The brewer should carefully avoid purchasing hops that are slack bagged, or kept in a moist place, or malt that has been sprinkled with water soon after it was taken from the kiln. By means of the moisture, an internal agitation is raised in the corn, which agitation, though soon stopped, for want of a sufficient quantity of air, yet, the heat thereby generated remaining, every adventitious seed, fallen from the air, and resting on the corn, begins to grow, and forms a moss, which dies, and leaves a putrid musty taste behind, always prevailing, more or less, in beer made from such grain.

That water is by no means an universal solvent, as some people have believed, has been already observed. It certainly does not act as such on metals, gems, stones, and many other substances: it is not in itself capable of dissolving oils, but is miscible with highly rectified spirits of wine, or alchohol, which is the purest vegetable oil in nature. All saponaceous bodies, whether artificial or natural, fixed or volatile, readily melt therein; and as many parts of the malt are dissoluble in it, they must either be, or become by heat, of the nature of soap, that is, equally miscible with oils and water.

When a saponaceous substance is dissolved in water, it lathers, froths, and bears a head; hence, in extracts of malt, we find these signs in the underback. Weak and slack liquors, which contain the salts of the malt without a sufficient quantity of the oils, yield no froth. Somewhat like this happens, when the water for the extract is over-heated, for then as more oils are extracted than are sufficient to balance the salts, the extract comes down as before, with little or no froth or head. This sameness of appearance, from two causes directly opposite to each other, has many times misled the artist, and shews the necessity there is to employ means less liable to error.

This might be a proper place to observe the difference between rain, spring, river, and pond waters; but as the art of brewing is very little affected by the difference of waters, if they be equally soft, but rather depends on the due regulation of heat; and as soft waters are found in most places, and become more alike, when heated to the degree necessary to form extracts from malt; it is evident, that any sort of beer or ale may be brewed with equal success, where malt and hops can be procured proper for the respective purposes. If hitherto prejudice and interest have appropriated to some places a reputation for particular sort of drinks, it has arose from hence; the principles of the art being totally unknown, the event depended on experience only, and lucky combinations were more frequent where the greatest practice was. Thus, for want of knowing the true reason of the different properties observed in the several drinks, the cause of their excellencies or defects was ignorantly attributed to the water made use of, and the inhabitants of particular places soon found an advantage, in availing themselves of this local reputation. But just and true principles, followed by as just a practice, must render the art more universal, and add dignity to the profession, by establishing the merit of our barley wines on knowledge, not on opinion void of judgment. To place this truth in a fuller light, and to communicate to the brewer the readiest means to examine any waters he may have occasion to use, I have extracted from Doctor Lucas's Essay on Waters, the experiments he made on the Thames, New River, and Hampstead company's waters, but without closely adhering to the accuracy this gentleman prescribed to himself; such exactness much better suiting a man of his abilities: for the purposes of brewing it is not of absolute necessity.

Experiments on the Thames, New River, and Hampstead Waters, which in general are in use in the Cities of London and Westminster.

Subjects employed.	Thames, at Somerset House.	Inferences from the experiments on Thames water.	New River.	Hampstead.
	Quantity of insoluble matter in one pint, one grain and a half.		Quantity of insoluble matter in one pint, one grain and a half.	In 24 hours discharges air, lets some light sediment fall, and grows clearer.
	Quantity of water used two ounces.		Quantity of water used two ounces.	Quantity of water used two ounces.
Infusion of campechy wood to a dark orange.	A pink color heighten to crimson.	A calcarious earth dissolved in a marine acid, perhaps something of a volatil alkaly, whence the water appears unfit for the scarlet dye.	A paler pink; but heightens as Thames.	A pink bloom; upon standing heightens; after fades, and comes to the color of old Canary Wine.
1 grain of cochinelle,	A pink bloom heightens	Confirms the preceding experiment.	The same as the Thames water.	A very beautiful crimson;

in powder.	to crimson; fades to a pale muddy purple, letting fall obscure green clouds.			heightens upon standing; in 12 hours suffers no diminution of color.
Alcaline lye, 5 drops.	Slight milky cloud; becomes milky all over; a light sediment of pale earth coats the glass, and is found at bottom.	Charged with terrine parts, dissolved by means of an acid; at high water more acid in the water than at low, and the alkaline principle in this river more at low water than at high.	Less milky, with less sediment.	Of alkaline lye used ten drops.-- Worked no sensible change in this water.
Solution of Soap.	A pearl-colored milkiness, but no coagulation.	Confirms the former observation.	Less milky; no coagulation.	Mixes smoothly, and causes a slight lactescence.
A diluted acid of vitriol.	No perceptible change.	Shews an alkaly not predominant.	No sensible change.	Upon standing shews some air bubbles, and seems somewhat brighter.
Mercury sublimate dissolved in pure water, 10 drops.	No change; upon standing, a mother of pearl colored pellicle	The quantity of alkaly inconsiderable.	The same appearance as Thames; rather slighter precipitation.	The same appearance, but rather slighter than any of

	covered the surface; the liquor beneath slightly milky.			the other two.
A solution of mercury in the acid of nitre.	Pale clouds at every drop; 1st white and milky, then yellowish four drops more got the same color all over; upon standing, a slight pale pellicle arose, and a muddy ochre-colored sediment subsided.	Shews some absorbent earth, by means of an acid, suspended in the water.	The same as Thames, but slighter.	Upon dropping, no change appears; upon standing grows milky, then to a pale yellow, with a slight pearl-colored pellicle; shews no air nor sediment; the glass slightly coated upon standing; precipitated fairly.
A solution of lead in distilled vinegar, at every drop as far as 4 drops.	A bright milky cloud, which, growing more opac and white, subsided; upon being stirred, had a milky opacity all	Confirms the preceding observation.	The same as Thames, but in a lower degree.	The same as New River.

	over; upon standing, threw up a pale pellicle, and let fall white precipitate.			
A solution of silver in the acid of nitre, 4 drops.	Caused a pearled milkiness; upon standing subsided a violet purple colored precipitate.	Shews some portion of sea-salt, of which the Thames has more at high water than at low.	The same effects, but slighter; the precipitate of a pale violet color.	Pale bluish white clouds; the precipitate, a bluish slate color, thinly covered the sides and bottom of the glass.

All these waters appear to be sufficiently pure for the common uses of life; the difference between them is very trivial, if any: those of Hampstead approach nearest to the simple state this element is to be wished for. Although it cannot be said to have an immediate relation to this work, yet it may not, perhaps, be disagreeable or useless here to add the quantities of water the cities of London and Westminster, and the adjacent buildings, are daily supplied with.

From the New River Company 57897 Tons per Day.

London Bridge,	8500	
Chelsea,	1740	
Hampstead,	1200	
York Buildings,	849	
Hartshorn Lane,	205	
	———	
	70391	Tons required every 24 hours.

SECTION IV.

OF EARTH.

REGULARITY requires some notice should be taken of this element. The great writer on chymistry, so often mentioned, defines it to be a simple, hard, friable, fossil body, fixed in the fire, but not melting in it, nor dissoluble in water, air, alcohol, or oil. These are the characters of pure earth, which, no more than any of the other elements, comes within our reach, free from admixture. Though it is one of the component parts of all vegetables, yet as, designedly, it is never made use of in brewing, except sometimes for the purpose of precipitation; it is unnecessary to say any thing more upon it: whoever desires to be farther informed concerning its properties may consult all, or any of the authors before mentioned.

SECTION V.

OF MENSTRUUMS OR DISSOLVENTS.

BY menstruums is understood a body which, in a fluid or subtilised state, is capable of interposing its small parts betwixt the small parts of other bodies. This act so obviously relates to the art of brewing, especially where the extracting of the malt and the boiling of the hops are concerned, that it should not be passed unheeded by.

The doctrine of menstruums, as laid down by Boerhaave, seems most intelligible and applicable to our purpose. He says, the solutions of bodies in general are the effect only of attraction and repulsion, between the particles of the menstruums and those of the body dissolved, the whole action depending on the relation between these two; of consequence, there cannot be any body, natural or artificial, which, without distinction, will dissolve all bodies whatsoever; nor is the cause assignable why certain menstruums dissolve certain bodies: the effects of alcaline, acid, neutral, fixed, or volatile salts, any more than those of oils, water, alcohol, fire, or air, are not to be accounted for by any general rule, that universally holds true; nor even, in many cases, doth the dissolution of a body depend on the purity or simplicity of the menstruum: the nearest path then to success, is cautiously to apply every menstruum we know of to the body whose solvent we want to discover.

The elements of fire and air greatly promote the action and effect of menstruums, and in this light they are admitted as such. Water dissolves most salts, all the natural sapos of plants, and the ripe juices of fruits; for in these, the oils, salts, and spirit of the vegetables, are accurately mixed and concreted together, and malts, having the same constituent parts with them, this element becomes a proper menstruum to extract this grain: though malts, by being dried with heats which greatly exceed what is necessary to bring barley to a state of maturity, do, from hence, require greater, though determinate heats, yet inferior to that at which water boils; but such heats must be applied in proportion to their dryness, to extract their necessary parts. Even earths, by the intervention of acids, dissolve in water; but having treated of the four elements already, as far as we conceived was requisite for the art of brewing, we shall, in this chapter, confine ourselves to oils and salts, and view these acting as menstruums only.

To the definition already given of oils, it may be necessary to add, in general, they contain some water, and a volatile acid salt; that they receive different appellations, and have different properties in proportion to their respective spissitudes. Oils from vegetables are obtained by expression, infusion, and distillation; in either of which methods, a too great heat is to be avoided, as this gives them a prejudicial rancidness, and where water does not interpose, alters their color until thereby they are turned black.

In general oils unite with themselves, but, excepting alcohol, not with water, unless when combined with salts, for salts attract water, and so they do oils: hence arises many elegant preparations both natural and artificial, from which wines are formed.

The power of oils in dissolving bodies is in a proportion to their heat, and being capable, when pure, of receiving a quantity of fire equal to 600 degrees, it is not surprising this liquid should mix with gums and with resinous bodies; but the color of these, and of every subject when thrown into boiling oils, changes in proportion to the impression made on them by heat, either to a yellow, a red, or a black. Oils which are inspissated, or thickened by heat, are termed balsams. Do not the oils of malt, from the heat they have undergone, resemble these? and from the circumstance of their having endured a heat superior to that necessary for putrefaction, may they not be suspected to possess a volatile alcaline salt? Beyond doubt, the extracts from malt (though they boil at a heat of 218 degrees only) yet do they, in great measure, dissolve hops, which are gum resinous.

Salt may well be denominated a menstruum, as it is easily diluted with water; fixed alcaline salts we have already seen appear to be the produce of fire alone.—Such are never distinguished in the composition of vegetables in their natural state; though a volatile alcalious salt (the effect of heat equal or superior to that necessary for putrefaction) is found in many, and especially in such as are putrified.

The power of a fixed alcali as a solvent is great, applied (says Boerhaave) to animal, vegetable, or fossil concretions, so far as they are oils, balsams, gummy, resinous, or of gummy resinous nature, and therefore concreted from oily substances: these, this salt intimately opens, attenuates, and resolves: disposing them to be perfectly miscible with water: oils of alcohol leaving however the impression of taste naturally belonging to this salt.

Vegetable acid salt dissolves animal, vegetable, fossil, and metalline substances, except mercury, silver, and gold. In most terrestrial vegetables this salt is evident; ripe mealy corn has the least indication of it, yet extracts therefrom, when fermented, and sometimes before they are fermented, discover sensibly their acidity. Sea-plants in general have not their roots inserted in the earth at the bottom of the sea, and these in distillation yield an oily volatile alcali; but more subtil than the native acids of vegetables, are the vinous acids produced by fermentation; they dissolve equally most matters put into them, and render the whole homogene. Into a must or wort, when under this act, by means of an elæosaccharum, might be introduced the choicest flavors, and the aromatics of the Indies be applied to heighten the taste and flavor of our barley wines. The laws of England at present subsisting are indeed opposite to any improvement of this sort, from the apprehensions of abuse: but where elegance alone is intended, undoubtedly the merit of our beers and ales might thereby be increased. As such, this is a part of chymical knowledge well worth the enquiry and attention of the brewer.

Neutral salts have already been mentioned; these are very various, and very different when acting as menstruums. Resins and gum-resins are generally said to be most effectually dissolved by alcohol; but Boerhaave informs us, that sal-amoniac (a very salutary subject and a neutral salt) if boiled with gums, resins, or the gum-resins of vegetables, intimately resolves, and disposes them to be conveniently mixed in aqueous and fermenting spiritous menstruums. Of this class of salts thus much is sufficient. This observation perhaps is of too much consequence to escape the notice of the artist.

SECTION VI.

OF THE THERMOMETER.

THIS instrument is designed for measuring the increase or decrease of heat. By doing it numerically, it fixes in our minds the quantity of fire, which any subject, at any time, is impregnated with. If different bodies are brought together, though each possesses a different degree of heat, it teaches us to discover what degree of heat they will arrive at when thoroughly mixed, supposing effervescence to produce no alteration in the mixture.

The inventor of this admirable instrument is not certainly known, though the merit of the discovery has been ascribed to several great men, of different nations, in order to do them and their countries honor. It came to us from Italy, about the beginning of the sixteenth century. The first inventors were far from bringing this instrument to its present degree of perfection. As it was not then hermetically sealed, the contained fluid was, at the same time, influenced by the weight of the air, and by the expansion of heat. The academy of Florence added this improvement to their thermometers, which soon made them more generally received; but, as the highest degree of heat of the instrument, constructed by the Florentine gentlemen, was fixed by the action of the strongest rays of the sun in their country, this vague determination, varying in almost every place, and the want of a fixed universal scale, rendered all the observations made with such thermometers of little use to us.

Boyle, Halley, Newton, and several other great men, thought this instrument highly worthy of their attention. They endeavoured to fix two invariable points to reckon from, and, by means of these, to establish a proper division. Monsieur des Amontons is said to have first made use of the degree of boiling water, for graduating his mercurial thermometers. Fahrenheit, indeed, found the pressure of the air, in its greatest latitude, would cause a variation of six degrees in that point; he therefore concluded, a thermometer made at the time when the air is in its middle state, might be sufficiently exact for almost every purpose. Long before the heat of boiling water was settled as a permanent degree, many means were proposed to determine another. The degree of temperature in a deep cave or cellar, where no external air could reach, was imagined by many a proper one; but what that degree truly was, and whether it was fixed and universal, was found too difficult to be determined. At last the freezing point of water was thought of, and though some doubts arose, with Dr. Halley and others, whether water constantly froze at the same degree of cold, Dr. Martine has since, by several experiments, proved this to be beyond all doubt, and this degree is now received for as fixed a point as that of boiling water.

These two degrees being thus determined, the next business was the division of the intermediate space on some scale, that could be generally received. Though there seemed to be no difficulty in this, philosophers of different countries have not been uniform in their determinations, and that which is used in the thermometer at present the most common, and, in other respects, the most perfect, is far from being the simplest.

The liquid wherewith thermometers were to be filled, became the object of another enquiry. Sir Isaac Newton employed, for this purpose, linseed oil; but this, being an unctuous body, is apt to adhere to the sides of the glass, and, when suddenly affected by cold, for want of the parts which thus stick to the sides, does not shew the true degree.

Tinged water was employed by others; but this freezing, when Fahrenheit's thermometer points 32 degrees, and boiling, when it rises to 212, was, from thence, incapable of denoting any more intense cold or heat.

Spirit of wine, which endures much cold without stagnating, was next made use of; but this liquor, being susceptible of no greater degree of heat than that which, in Fahrenheit's scale, is expressed by 175, could be of no service where boiling water was concerned.

At last the properest fluid, to answer every purpose, was found to be mercury. This had never been known to freeze[6]; and not to boil under a heat of 600 degrees, and is free from every inconveniency attending other liquors.

As the instrument is entirely founded on this principle, that heat or fire expands all bodies, as cold condenses them, there was a necessity of employing a fluid easy to be dilated. A quantity of it is seated in one part in the bulb. This being expanded by heat, is pushed forward into a fine tube, or capillary cylinder, so small, that the motion of the fluid in it is speedy and perceptible. Some thermometers have been constructed with their reservoir composed of a larger cylinder; but in general, at present, they are made globular. The smaller the bulb is, the sooner it is heated through, and the finer the tube, the greater will be the length of it, and the more distinct the degrees. It is scarcely possible that any glass cylinder, so very small, should be perfectly regular; the quicksilver, during the expansion, passing through some parts of the tube wider than others, the degrees will be shorter in the first case, and longer in the latter. If the divisions, therefore, are made equal between the boiling and freezing points, a thermometer, whose cylinder is irregular, cannot be true. To rectify this inconveniency, the ingenious Mr. Bird, of London, puts into the tube about the length of an inch of mercury; and measuring, with a pair of compasses, the true extent of this body of quicksilver in one place, he moves it from one end to the other, carefully observing where it increases or diminishes in length, thereby ascertaining the parts, and how much the degrees are to be varied. By this contrivance, his thermometers are perfectly accurate, and exceed all that were ever made before.

I shall not trouble my reader with numerous calculations that have been made, to express the quantity of particles of the liquor contained in the bulb, in order to determine how much it is dilated. This, Dr. Martine seems to think a more curious than useful enquiry. It is sufficient, for our purpose, to know how the best thermometers ought to be constructed: they who have leisure and inclination, may be agreeably entertained by the author last cited.

By observing the rise of the mercury in the thermometer, during any given time, as, for instance, during the time of the day, we ascertain the degree and value of the heat of every part of the day, from whence may be fixed the medium of the whole time, or any part thereof. By repeated experiments, it appears, the medium heat of most days is usually indicated at eight o'clock in the morning, if the instrument is placed in the shade, in a northern situation, and out of the reach of any accidental heat.

Though water is not so readily affected as air by heat and cold, yet, as all bodies long exposed in the same place, become of the same degree of heat with the air itself, no great error can arise from estimating water, in general, to be of the same heat as the air, at eight o'clock in the morning, in the shade.

The thermometer teaches us that the heat of boiling water is equal to 212 degrees, and by calculation we may know what quantity of cold water is necessary to bring it to any degree we choose; so, notwithstanding the instrument cannot be used in large vessels, where the water is heating, yet, by the power of numbers, the heat may be ascertained with the greatest accuracy. The rule is this: multiply 212, the heat of boiling water, by the number of barrels of water thus heated, (suppose 22) and the number of barrels of cold water to be added to the former, (suppose 10,) by the heat of the air at eight o'clock, (suppose 50,) add these two products together, and divide by the sum of the barrels; the quotient shews the degree of heat of the water mixed together.

	212	heat of boiling water.	
	22	barrels to be made to boil.	
	—		
	424	50 deg. heat of air at eight.	

		424	10 barrels of cold water.
		—	—
	22	4664	500
	10	500	
	—	—	
sum 32) of barrels 32		5164	(161⅓ degrees will be the heat of the water when mixed together.
		—	
		196	
		192	
		—	
		44	
		32	
		—	
		12	

The calculation may be extended to three or more bodies, provided they be brought to the same denomination. Suppose 32 barrels of water to be used where there is a grist of 20 quarters of malt, if these 20 quarters of malt are of a volume or bulk equal to 11 barrels of water, and the malt, by having lain exposed to the air, is of the same degree of heat with the air, in order to know the heat of the mash, the calculation must be thus continued.

	161⅓	heat of water 50 degrees of heat of malt
	32	barrels of water 11 barrels, volume of malt
	—	—
	333	550
	483	
	—	
32 water	5163	

11 malt	550	
43	) 5713	(132 degrees, which will be the heat of the mash.
	43	
	141	
	129	
	123	
	86	
	37	

We shall meet hereafter with some incidents, which occasion a difference in the calculations made for the purpose of brewing, but of these particular mention will be made in the practical part.

The thermometer, by shewing the different degrees of heat of each part of the year, informs us, at the same time, how necessary it is the proportions of boiling water to cold should be varied to effect an uniform intent; also that the heat of the extracts of small beer should differ proportionably as the heats of the seasons do: it assists us to fix the quantity of hops necessary to be used at different times; how much yeast is requisite, in each term of the year, to carry on a due fermentation; and what variation is to be made in the length of time that worts ought to boil. Indeed, without this knowledge, beers, though brewed in their due season, cannot be regularly fermented, and whenever they prove good, so often may it be said fortune was on the brewer's side.

Beers are deposited in cellars, to prevent their being affected by the variations of heat and cold in the external air. By means of the thermometer, may be determined the heat of these cellars, the temper the liquor is kept in, and whether it will sooner or later come forward.

The brewing season, and the reason why such season is fittest for brewing, can only be discovered by this instrument. It points out likewise our chance for success, when necessity obliges us to brew in the summer months.

As all vegetable fermentation is carried on in heats, between two settled points, we are, by this instrument, taught to put our worts together at such a temperature, as they shall neither be evaporated by too great a heat, nor retarded by too much cold.

If curiosity should lead us so far, we might likewise determine, by it, the particular strength of each wort, or of every mash; for if water boils at 212 degrees, oil at 600, and worts be a composition of water, oil and salt; the more the heat of a boiling wort exceeds that of boiling water, the more oils and salts must it contain, or the stronger is the wort.

A given quantity of hops, boiled in a given quantity of water, must have a similar effect, consequently the intrinsic value of this vegetable may, in the same manner, be ascertained.

The more the malts are dried, the more do they alter in color, from a white to a light yellow, next to an amber, farther on to a brown, until the color becomes speckled with black; in which state we frequently see it. If more fire or heat is continued, the grain will at last charr, and become intirely black. By observing the degrees of heat necessary to induce these alterations, we may, by the mere inspection of the malt, know with what degree of fire it has been dried; and fixing upon such which best suits our purpose, direct, with the greatest accuracy, not only the heat of the first mash, but the mean heat the whole brewing should be impressed with to answer our intent, circumstances of the greatest consequence to the right management of the process.

If I had not already said enough to convince the brewer of the utility of this instrument, how curious he ought to be in the choice, and how well acquainted with the use of it, I should add the heat gained by the effervescing of malt, is to be determined by it alone; the quantity of heat lost by mashing, by the water in its passage from the copper to the mash ton, and by the extract coming down into the underback, these can be found by no other method; and, above all, that there is no other means to know with certainty the heat of every extract.

I know very well good beers were sometimes, perhaps often, made before the thermometer was known, and still is, by many who are entirely ignorant of it; but this, if not wholly the effect of chance, cannot be said to be very distant from it. They who carry on this process, unassisted by principles and the use of the thermometer, must admit they are frequently unsuccessful, whereas did they carefully and with knowledge apply this instrument, they certainly would not be disappointed.—It is equally true, the brewing art, for a long space of time, has been governed by an ill-conveyed tradition alone; if lucky combinations have sometimes flattered the best practitioners, faulty drinks have as often made them feel the want of certain and well established rules. It is just as absurd for a brewer to refuse the use of the thermometer, as it would be for an architect to reject the informations of his plummet and rule, and to assert they were unserviceable because the first house, and probably many others, were built without their assistance.

SECTION VII.

OF THE VINE, ITS FRUITS, AND JUICES.

AFTER these short accounts of the principles and instrument necessary to the right understanding of the brewing art, we should now draw near to the particular object of this treatise, but as the most successful method to investigate it, must be first to inspect the great and similar example nature has set before us, our time will not be lost by making this enquiry.

Any fermented liquor, that, in distillation, yields an inflammable spirit miscible with water, may be called wine, whatever vegetable matter it is produced from.—As beer and ales contain a spirit exactly answerable to this definition, brewing may justly be called the art of making wines from corn. Those, indeed, which are the produce of the grape, have a particular claim to the name, either because they are the most ancient and the most universal, or that a great part of their previous preparation is owing to the care of nature itself. By observing the agents she employs, and the circumstances under which she acts, we shall find ourselves enabled to follow her steps, and to imitate her operations.

Most grapes contain juices, which, when fermented, become in time as light and pellucid as water, and are possessed of fine spiritous parts, sufficient to cherish, comfort, and even inebriate. But these properties of vinosity are observed not to be equally perfect in the fruits of all vines; some of them are found less, others not at all proper for this purpose. It is therefore necessary to examine the circumstances which attend the forming and ripening of those grapes, whose juices produce the finest liquors of the kind.

All grapes, when they first bud forth, are austere and sour, therefore of a middle nature. And this can be no other than the effect of the autumnal remaining sap, mixed with the new raised vernal one, the consequence of which mixture will be found greatly to merit our inquiry. As far as our senses can judge, at first, it appears that the juice, in this state, consists of somewhat more than an acid combined with a tasteless water. When the fruit is ripe, it becomes full of a rich, sweet, and highly flavoured juice. The color, consistency, and taste of which shew, that, by the power of heat, a considerable quantity of oil has been raised, and, sheathing the salts, is the reason of its saccharine taste and saccharine properties.

In England, grapes are probably produced under the least heat they can be raised by. They discover themselves in their first shape, about June, when the medium heat of the twenty-four hour's shade is 57,60. This, with what more should be added for the effect of the sun's beams, are the degrees of heat which first introduce the juices into this fruit.

The highest degrees of heat, in the countries where grapes come to perfect maturity, have been observed to be, in various parts of Italy, Spain, and Greece 100, and at Montpelier 88, in the shade; to which, according to Dr. Lining's observations, 20 degrees must be added for the effect of the sun's beams. The greatest heat in Italy will then amount to 120 degrees, and in the south of France to 108. These approach nearly to the strongest heats observed in the hottest climates, which, in Astracan, Syria, Senegal, and Carolina, were from 124 to 126 degrees.

Those countries, where the heat is greatest, in general produce the richest fruits, that is, the most impregnated with sweet, thick and oily juices. We are told, among the Tockay wine-hills, there is one which, directly fronting the south, and being the most exposed to the sun, yields the sweetest and richest grapes. It is called the *sugar-hill*, and the delicious wines extracted from this particular spot, are all deposited in the cellars of the imperial family. Those grapes, some in the Canaries, some in other places, being suffered to remain the longest on the tree, with their stems half cut through, by this means procure their juices to be highly concentrated, and produce that species of sweet, oily, balmy wines, which, from this operation, are called *sack*, a derivation of the French word *sec* or *dry*.

In all distillations of unfermented vegetables, water and acid salts rise first. A more considerable degree of fire is required for the elevation of oils, and a still greater one for the lixivial salts, which render those oils miscible with water.

A plant, exposed to a very gentle heat, at first yields a water which contains the perfect smell of the vegetable blended with a subtile oil; if more heat be added, an heavier oil will come over: from some a volatile alkali, from others a phlegm will rise, which gradually grows acid; and, last of all, with the farther assistance of fire, the black, thick, empyreumatic sulphur. Nature, in a less degree, may be said to place a like series of events before our eyes, in the forming and maturating of grapes, and it is by imitating what she does, that the inhabitants of different countries may improve the advantages of their soil and of their air.

In order to illustrate the doctrine, that grapes are endued with various properties, in proportion to the heat of the air they have been exposed to, let us remember what Boerhaave has observed, that, in very hot weather, the oleous corpuscles of the earth are carried up into the air, and, descending again, cause the showers and dews in summer to be very different from the pure snow of winter. The first are acrid, and disposed to froth, the last is transparent and insipid. Hence summer rain, or rain falling in hot seasons, is always fruitful, whereas in cold weather it is scarcely so at all. In winter the air abounds with acid parts, neither smoothed by oils nor rarified by heat: cold is the condensing power, as heat is the opener of nature. In summer, the air, dilating itself, penetrates every where, and gives to the rain a disposition to froth, occasioned by the admixture of oleous and aërial particles. Thus the acid salts, either previously existing, or by the vernal heat introduced into the grapes, and necessary to their preservation, are neutralized by coming in contact with the juices the foregoing autumn produced; after which a hotter sun, covering or blending these juices with oils, changes the whole into a saccharine form. In proportion as these acids are more or less sharp, and counterbalanced by a greater or lesser quantity of oils, the juices of grapes approach more or less to the state of perfection, which fermentation requires.

There are many places, as Jamaica, Barbadoes, &c. in which experience shews the vine cannot be cultivated to advantage. By comparing the heat of these places with those in Italy and Montpelier, it appears this defect is not owing to excessive heats, but to their constancy and uniformity; the temperature of the air of these countries seldom being so low as the degree necessary for the first production of the fruit. Whenever the cultivation of the vine is attempted in these parts of the West Indies, the grapes, on their first appearance, are shaded and skreened from the beams of the sun, which, in their infancy, they are not able to bear.

Hence we learn, though nature employs both the autumnal and vernal seasons, yet there are lesser heats with which she prepares the first juice of grapes, a stronger power of the sun she requires to form the fruit, and a greater than either to ripen it. We have investigated the lowest degrees of heat, in which grapes are produced, and nearly the highest they ever receive to ripen them. Let us call the first the *germinating* degrees, and the last those of *maturation*. If nearly 58 be the lowest of the one, and 126 the highest of the other, and if a certain power of acids is necessary for the germination of the grapes, which must be counterbalanced by an equal power of oils raised by the heat of the sun for their maturation, then the medium of these two numbers, or 92, maybe said to be a degree at which this fruit cannot possibly be produced, and inferior to that by which it should be matured. At Panama the lowest degree of heat in the shade is 72, to which 20 being added, for the sun's beams, the sum will be 92, and consequently no grapes can grow there, except the vines be placed in the shade.

If we recollect that we can scarcely make wine, which will preserve itself, of grapes produced in England, we shall be induced to think, that the reason of this defect is the want of the high degrees of heat. Our sun seldom raises the thermometer to 100 degrees, and that but for a short continuance. Our medium heat is far inferior to 92, and hence we see, at several distant terms in summer, new germinated grapes, but seldom any perfectly ripe. These observations, the use of which, in brewing, we will endeavour to apply, likewise point out to us, what part of our plantations are fit to produce this fruit, and to what degree of perfection.

A research made for each constituent part forming grapes, as well as the proportion they bear to one another, at first sight, appears to be an eligible method to discover the nature of wines; but in every vegetable their parts are mixed and interwoven, and every degree of heat, acting on them, finds these so blended, as to render their division too imperfect for such enquiry to be made with sufficient accuracy, to deduce therefrom the rules of an art. In the producing, ripening, and fermenting the juice of the grapes, as well as in forming beers and ales, the element of fire so superlatively influences and governs every progressive act, as to occasion some remarkable difference in their appearance: from, hence, then, we may expect the information we want, and be enabled to discover the laws by which Nature forms her wines.

When the constituent parts of a subject are to be estimated by heat alone, the number of degrees comprehended between the first heat which formed it, and the last which brought it to a perfect state, must express the whole of its constituent parts. Complete finished substances, must have been benefited by the whole latitude of degrees applicable thereto; and in proportion as part of the whole latitude is wanting, will their nature be different, and themselves less perfect.

This variety is remarkable in the fruit we are now treating of. A country endued with the lowest germinating, and with the highest maturating degrees of heat for grapes, would produce them in the utmost perfection; that is, they would possess all the several properties they could obtain from this circumstance; consequently such are capable of forming wines that would preserve themselves a very long time, and would also become spontaneously fine. From the several heats we have observed that this fruit is capable of enduring, it is reasonable to believe the greatest number of degrees of heat employed to form all their constituent parts, must be where, during the whole space of vegetation, the heat in the shade varies from 60 to 106 degrees, and constitutes a difference of 46 degrees. So great a latitude, ordered by nature, most certainly denotes the general utility of the plant.

The climate of the southern part of France approaches nearest to this; but Spanish wines are richer; their grapes are formed by a warmer sun; their vernal and maturating heats exceed those of France; but, at the same time, their wines are more stubborn, and, to be made fine, require the help of precipitation. This variety increases according to the heat of climates: thus we see wines which come from the coast of Africa, whose richness and stubbornness are beyond the reach of any menstruum employed to fine them. Let us endeavour to reduce this apparent inconstancy to rule, in order to assist our art.—If the lowest heat which forms the grape, in the southern parts of France, be 60 degrees, and if 88 degrees, in the shade, be the mean of their maturating heat, the difference between 60 and 88, or 28 degrees, is the number which includes the constituent parts of grapes in this country, as these degrees imply the whole space of their progress. If like juices were to be imitated by art, as in our hot-houses, it is clear half the number of the degrees of heat which form the whole of the constituent parts, or 14, deducted from 74, the mean heat of their whole vegetation, would give 60, for the first heat to be employed, and this to be raised, for maturation, to 88, the greatest heat, nature in this case, permits, or 14 degrees to be added to the same whole mean. To liken the wines of Spain, where the autumnal and vernal heats are greater than in France, the heat forming the first juices must be more, as also the maturating heats; but with such practice, the number of constituent degrees would be found to be fewer, and spontaneous brightness could no more be expected, than it is found, in their wines.

A strict enquiry after the heats first and last applied to grapes, is of such consequence to ascertain the principles by which malt liquor should be formed, that, though grapes produced in England scarcely make wines which can maintain themselves sound, yet, as the rule is universal, even from them we shall be able to establish not only its certainty, but also the application of the number of the degrees found between the heats which germinate the fruit, and those which ripen them.

	Deg.
From twelve years observation, we have found the mean heat in the shade, from the 1st of June, to the 15th, when grapes with us first bud forth, to be	57.60
Our greatest heat, under like circumstances, from the 15th to the 31st of July, to be	61.10
	———
Their difference,	3.50
	———
Their medium,	59.35
	———

If, from their medium, 59.35, we subtract 1.75, half their difference, or half their constituent parts, we must have left 57.60 for the germinating heat; and if to their medium, 59.35, we add 1.75, half the number of their constituent parts, we shall have 61.10, the highest mean heat, in the shade, at the time the richest juices of our grapes are formed. It is true, in July, nor even in the following months, when the heat continues nearly alike, our grapes are not ripe, nor gathered; the properties raised by our greatest sunshine, as yet have not reached the fruit, and though the mean heat of the air in September and October is less, yet it is sufficient to place in the grapes the juices raised by the preceding hot sun, which concentrate and grow richer, by remaining on the plant, though, for want of a sufficient heat, they do not reach that perfection obtained in warmer climates.

The want of grapes in many parts both of America and Africa, and the reason we gave for this, (See page 55,) warrants the truth of the division we have just now made, between the germinating and maturating heats; and if the effects caused by a hot sun do not immediately benefit the fruit, by a parity of reason, after the grapes are gathered, the plant must possess, (and surely for some longer space, by a continued heat, equal, and often superior, to the vernal sun,) juices which Nature is too frugal not usefully to apply; these juices, we apprehend, assist in forming the embryo of the leaves which are fully to expand the ensuing year, and serve, by their oleaginous quality, to preserve these and the whole plant during the cold of the winter; which cold, at the same time that it contracts the pores of the vine, condenses and thickens these richer juices, from whence few, if any of them, are lost or expended by perspiration. The heat of the following spring renews their activity, when blending with those this season attracts, the leaves open, the flowers appear, and the fruit forms. Thus far we conceive the act of germination extends, provided for and assisted both by the autumnal and vernal heats, and which, in point of power, are nearly equal and uniform.

The heat of the sun, during summer months, and if to this we add the more constant heat at the roots of the vine, retained there by the density of the earth; these (though superior to the germinating heat) produce a like uniformity for maturating the fruit: thus nature, in order to implant in wines an original even taste, and to facilitate the fermentable act, amidst the great variety that appears to us in the heat of the air, seems, upon the whole, to act by steady and equal motions; or rather, perhaps, this is the best manner by which we can reduce to rule; the inconstancy of the atmosphere.

I am sensible these facts had been represented in a more natural light, had I observed the degrees of heat impressed on the vine in every season of the year; the difference of the sun's heat, in every hour of the day, a variety exceeding that in the shade; that between night and day; the aspect of the plant; the heat of the earth at its surface, as well as at the roots of the vine; all these would have increased the circumstances to a prodigious extent; which, though perhaps requisite to satisfy philosophic investigation, might, from their number and variety, have been the means rather to induce us to error, than to discover the general rules by which nature acts.

From the above-related process we are taught, that nature, in forming wines, is not confined to a certain fixed number of degrees, but admits, for this act, of a considerable latitude, according to the extent of which the wines vary in taste and properties; and that she affects an equality of heat in each period of vegetation; from whence the brewer is taught, if he form his malt-liquors with four mashes, as in the autumn and spring the vine is impressed with heats nearly uniform, so ought his two first mashes to be; the third, in imitation of the high heat of summer, should be much hotter, and the heat of his last mash the same with this; and this general rule has been found universally true, for beers expected to preserve themselves sound a sufficient time; and admits but of a proportional variation, when fewer or more mashes are employed, as the degrees of heat denominating the constituent parts of the grain, must be applied in proportion to the quantity of water used to each mash; but in malt liquors speedily to be drank, or when we deviate greatly from the more perfect productions of nature, we are then compelled to swerve from her rules; a practice never profitable, and which nothing but necessity can justify.

The nature of the soil proper for the vine, might, in another work, be a very useful enquiry. It will be sufficient here, barely to hint at the effect, which lixivial soils produce in musts. The Portugueze, when they discovered the Island of Madeira in 1420, set fire to the forests, with which it was totally covered. It continued to burn for the space of seven years, after which the land was found extremely fruitful, and yielding such wines, as, at present, we have from thence, though in greater plenty. It is very difficult to fine these wines, and, though the climate of this island is more temperate than that of the Canaries, the wines are obliged to be carried to the Indies and the warmer parts of the globe, to be purged, shook, and attenuated, before they can arrive to an equal degree of fineness with other wines; were the Portugueze acquainted with what may be termed the artificial method of exciting periodical fermentation, much or the whole of this trouble might be avoided. Hence we see, that soils impregnated with alkaline salts will produce musts able to support themselves longer, and to resist acidity more, than other soils, under the same degree of heat.

Grapes have the same constituent parts as other vegetables. The difference between them, as to their tastes and properties, consists in the parts being mixed in different proportions. This arises, either from their absorbent vessels more readily attracting some juices than others, or from their preparing them otherwise, under different heats and in different soils.

We find, says Dr. Hales, by the chymical analysis of vegetables, that their substance is composed of sulphur, volatile salts, water, and earth, which principles are endued with mutual attracting powers. There enters likewise in the composition, a large portion of air, which has a wonderful property of attracting in a fixed, or of repelling in an elastic state, with a power superior to vast compressing forces. It is by the infinite combinations, actions, and reactions of these principles, that all the operations in animal and vegetable bodies are effected.—Boerhaave, who is somewhat more particular with regard to the constituent parts of vegetables, says, that they contain an oil mixed with a salt in form of a sapo, and that a saponaceous juice arises from the mixture of water with the former.

Thus we see, from the composition of grapes, that they have all the necessary principles to form a most exquisite liquor, capable, by a gentle heat, to be greatly attenuated. They abound with elastic air, water, oils, acid, and neutral salts, and even saponaceous juices.—The air contained in the interstices of fluids is more in quantity than is commonly apprehended. Sir Isaac Newton has proved that water has forty times more pores than solid parts; and the proportion, likely, is not very different in vegetable juices. When the fruit is in its natural entire state, the viscidity of the juices, and their being enveloped by an outward skin, prevent the expansion of the inclosed air; it lies as it were inactive. In this forced state, it causes no visible motion, nor are the principles, thus confined, either subjected to any apparent impressions of the external atmosphere, or so intimately blended as when they are expressed. A free communication of the external air, with that contained in the interstices of the liquor, is required to form a perfect mixture. By what means this is effected, what alterations it produces, or, in general, in what manner the juice of the grape becomes wines, must be the subject of our next inquiry.

The process of a perfect fermentation is undoubtedly the same (where the due proportions of the constituent parts, forming the must, are exactly kept) whatever vegetable juices it is excited in. For this reason, we will observe the progress of this act in beers and ales, these being subjects we are more accustomed to, and where the characters appear more distinct, in order to apply what may be learned from thence to our chief object, the business of the brewer.

SECTION VIII.

OF FERMENTATION IN GENERAL.

VEGETABLE fermentation is that act, by which oils and earth, naturally tenacious, by the interposition of salts and heats, are so much attenuated and divided, as to be made miscible with, and to be suspended in, an homogeneous pellucid fluid; which, by a due proportion of the different principles, is preserved from precipitation and evaporation. According to Boerhaave, a less heat than forty degrees leaves the mass in an inert state, and the particles fall to the bottom in proportion to their gravity; a greater heat than eighty degrees disperses them too much, and leaves the residuum a rancid, acrimonious, putrid mass.

It is certainly very difficult, if not impossible, to discover the true and adequate cause of fermentation. But, by tracing its several stages, circumstances, and effects, we may perhaps perceive the agents and means employed by nature to produce this singular change; a degree of knowledge, which, we hope, is sufficient to answer our practical purposes.

The must, when just pressed from the grapes, is a liquid, composed of neutral and lixivial salts, oils of different spissitude, water, earth, and elastic air. These, irregularly ranged, if I may be permitted the expression, compose a chaos of wine. Soon after the liquor is settled, a number of air bubbles arise, and at first adhere to the sides of the containing vessel; their magnitude increases as they augment in number, so that at last they cover the whole surface of the must.

It has been long suspected, and, if I mistake not, demonstrated, that an acid, of which all others are but so many different species, is universally dispersed through, and continually circulating in, the air; and that this is one of nature's principal agents, in maturating and resolving of bodies. Musts, like other bodies, being porous, the circulating acids very powerfully introduce themselves therein by the pressure of the atmosphere, in proportion as the pores are more or less expanded by the heat they are exposed to. The particles of acids are supposed by Newton to be endued with a great attractive force, in which their activity consists. By this force, they rush towards other bodies, put the fluid in motion, excite heat, and violently separate some particles in such manner as to generate or expel air, and consequently bubbles.

From hence it appears that, as soon as the acid particles of the air are admitted into the must, they act on the oils, and excite a motion somewhat like the effervescence generated, when acids and oils come in contact, though in a less degree. This motion is the cause of heat, by which the included elastic air, being rarefied, occasions the bubbles to ascend towards the surface.— These, by the power of attraction, are drawn to the sides of the vessel; at first they are small and few, but increase, both in number and magnitude, as the effect of the air continues, till, at last, they spread over the whole surface. The first stage of vegetable fermentation shews itself to be a motion excited by the acids floating in the air, acting on the oleous parts of the liquor, which motion gives an opportunity to the divided minute parts of air, dispersed throughout the whole, to collect themselves in masses: from hence they become capable to exert their elasticity, and to free themselves from the must. (See Arbuthnot on air p. 116.) It may, perhaps, be proper to observe, that all musts, which ferment spontaneously, contain for this purpose a large portion of elastic air.

Bubbles still continue to rise after the must is entirely covered with them; and a body of bladders is formed, called, by the brewers, the *head of the drink*; as the bubbles increase, the head rises in height, but the oils of the must, being as yet of different spissitudes, those which are least tenacious soon emit their air; others, somewhat stronger, being rarefied by the fermenting heat, rise on the surface higher than the rest, while such aerial bubbles as are more dense, take their place below them. From hence, and from the constituent parts of the drink not being as yet intimately mixed, the head takes an uneven and irregular shape, and appears like a beautiful piece of rock work. After this, it requires some time, and it is by degrees, that the particles dispose themselves in their due order, to be farther attenuated by the act of fermentation, which, when effected, the saline, oily, and spiritous parts become perfectly miscible with the water. The head of the liquor then is more level; heterogeneous bodies, as dirt, straw, corks, &c. assisted by bubbles of air adhering to them, are now buoyed on the surface, and should be skimmed off, lest, when the liquor becomes more light and spiritous, they should subside. About this time, such parts of the must as are too course to be absorbed in the wine (as they consist chiefly of pinguious oils, mixed with earth, though they strongly envelope much elastic air) from their weight, sink to the bottom, and form the lees. But the internal motion increasing, the air bubbles grow larger; some, not formed of parts so strong as the others, which generally are the first, burst and strengthen the rest; and thereby a heat is retained in the fermenting liquor, which carries the act on to a farther degree. The particles of the must become more pungent and spiritous, because more fine and more active; some of the most volatile ones fly off; hence, that subtle and dangerous vapor, called *gas*, which extinguishes flame and suffocates animals. The wine, by these repeated acts, being greatly attenuated, is at last unable to support, on its surface, the weight of such a quantity of froth, rendered more dense by the repeated explosions of the air bubbles. Now, lest the liquor should be fouled by the falling in of the froth, it is put in vessels having only a small aperture, where it continues to ferment, with a slower and less perceptible motion, which gradually diminishing until it reaches the period when it neither attracts or repels air, it admits of its communication with it to be cut off; not that thereby, in a strict sense, the fermentation can be said to be completely ended: the least heat is sufficient to renew, or rather to continue the act, more especially if by any means the atmosphere can gain any admittance, however small.

The alteration caused in the liquor, by the pressure of the external air, from the very first of its fermenting, not only occasions the particles of the must to form themselves in their due order, but also, by the weight and action of that element, grinds and reduces them into smaller parts. From hence they more intimately blend with each other, the wine becomes of an equal and even taste, and if the constituent parts of the must be in a perfect proportion, it will continue to ferment, until, these being disposed and ranged in right lines, a fine and pellucid fluid is produced.

That this operation subsists, even after the liquor becomes fine, is evident; for every fretting is a continuance of fermentation, though often almost imperceptible. Thus, the component parts of the liquor are continually reduced to a less volume, the oils become more attenuated, and less capable of retaining elastic air. As these frettings are often repeated, it is impossible to determine, by any rule, the exact state in which wine should be, in order to be perfect for use. It would seem, however, that the more minutely the parts are reduced, the more their pungency will appear, and the easier their passage be in the human frame. Both wines and beers, when new, possess more elastic air, than when meliorated by age; to be wholesome, they must be possessed of the whole of the fermentable principles. For these reasons, beers and ales, when substituted for wines in common, and more especially when given to the sick, should always be brewed from entire malt: for the last extracts, possessing but the inferior virtues of the grain, have by so much less the power to become light, spiritous, and transparent.

Wines never totally remain inactive; fermentation in some degree continues, and in time the oils, by being greatly attenuated, volatilise, fly off, and permit a readier admission of the external air into the drink. In proportion as this circumstance takes place, the latent acids of the liquor shew themselves, the wine becomes sour, and in this state is termed vinegar.

Its last stage or termination is, when the remaining active principles, which the vinegar possessed, being evaporated in the air, a pellicle forms itself on the surface of the liquor, and dust and seeds, which always float in the atmosphere, depositing themselves thereon, strengthen this film into a crust, on which grows moss, and many other small plants. These vegetables, together with the air, exhaust the watery parts; after which no signs of fermentable principles remain but, like the rest of created beings, all their virtues being lost, what is left is a substance resembling common earth.

Upon the whole, then, it appears, that a liquor fit for fermentation must be composed of water, acids smoothed over with oils, or saccharine salts, and a certain portion of elastic air; the heat of the air the liquor is fermented in, must be in proportion to the density of its oils; and lastly, that the pores are to be expanded by slow degrees, lest the air, by being admitted too hastily, should cause an effervescence rather than a fermentation, and occasion the whole to become sour. Wines, therefore, fermented in countries where the autumn is hot, require their oils to be more pinguious, than where the season is cooler. For the same reason beers are best made, when the air is at forty degrees of heat, or below the first fermentable point, because the brewer, in this case, can put his wort to work, at a heat of his own chusing, which will not be increased by that of the air; on the contrary, when, by its internal motion, the heat becomes greater, it will again be abated and regulated by the cold of the medium.

The pores of a wort are expanded in proportion to the heat it is impressed with; on which account common small beer, brewed in summer, when the air and acids more easily insinuate themselves into the liquor, ought to be enriched with oils obtained by hotter extracts, to sheath these salts; and in winter the contrary method must be pursued.

From this history of fermentation, we can, with propriety, account for the many accidents and varieties that accompany this act; and a comparative review of some of them may not be unnecessary.

A cold air, closing the pores of the liquor, always retards, and sometimes stops, fermentation; heat, on the contrary, constantly forwards this act; but, if carried too high, immediately prevents it.

A must, loaded with oils, will ferment with more difficulty than one which abounds with acids; it likewise is longer before it becomes perfectly fine; but, when once so, will be more lasting.

If the quantities of oil are increased, they will exceed the power both of the acids naturally contained in the must, and of those absorbed from the air in fermenting; the liquor will therefore require a longer time before it becomes pellucid, unless assisted by precipitation: and there may be cases where even precipitation cannot fine it.

These considerations naturally lead us to a general division of wines into three classes: First, of such as soon grow fine, and soon become acid, being the growth of cold countries. Secondly, of those which, by a due proportion of heat, both when the grapes germinate, and when they come to maturity, form a perfect must; and not only preserve themselves, but, in due time, (more especially when assisted by precipitation,) become transparent; and, thirdly, of such as, having taken their first form under the highest degrees of *germination*, (as I termed them) are replete with oils, disappoint the cooper, and render the application of menstruums useless, unless in such quantities as to change the very nature of the wine.

This remarkable difference in wines appears chiefly to arise from the climate; and it will confirm the observation before made, that, as wines are neither naturally nor uniformly perfect, they must be subject to many diseases.

All vegetable substances possess fermentable principles, though in a diversity of proportions; for those juices only, whose constituent parts approach to the proportion necessary for the act of fermentation, can be made into wines. I would not, however, from what I have attributed to a difference of heat in different climates, be understood, as if I thought that vegetables are more or less acid, more or less sulphureous, or in general more or less fermentable, merely from the heat of the country they grow in. This, though likely one of the principal causes of their being so, is by no means the only one; the form and constitution of the plant is another. In very hot climates, we find acid fruits, such as limes, tamarinds, lemons, and oranges; the proportions of fermentable principles in these fruits are such, as to render them incapable of making sound wines, though their juices may, in some degree, be susceptible of fermentation. In countries greatly favored by the sun, some vines and other fruit trees there are, which attract the acids from the air, and possibly from the earth, so greedily, that, when their juices are fermented, they soon become sour. On the contrary, in cold climates, we see warm aromatic vegetables grow, as hops, horse-raddish, camomile, wormwood, &c. whose principles cannot, without difficulty, and perhaps not perceptibly, be brought to ferment. But these instances must be accounted the extremes on each sides; for in cold, as well as in hot countries, fruits are produced susceptible of a perfect natural fermentation: with us, for example, apples; some species of which are endued with such austere and aromatic qualities, that their expressed juices ferment spontaneously, until they become pellucid, and are capable of remaining in a sound state many years. From hence it appears, that proper subjects, which will naturally ferment, for making wines, may be found in almost every climate. England, says Boerhaave, on this account, is remarkably happy: her fruits are capable of producing a great variety of wines, equal in goodness to many imported, were not our tastes but too often subservient, not to reason, but to custom and prejudice.

A similar want of perfection to that observed in wines, may be noticed in our beers and ales, and it chiefly has its origin in the different degrees of heat the malt has been impressed with, both in drying and extracting; where, in the processes of malting and brewing, a sufficient heat has not been maintained, the liquor undoubtedly must become acid; in proportion as the contrary is the case, or that the beer is overcharged with hops, if this is in no great excess, it retains still a greater tendency to fermentation than to putrefaction, acids not being wanting, but only enveloped. In this case, time will get the better of the disease; like to the wines made from the growth of too hot a sun, these liquors, at a certain period, sicken, smell rancid, and have a disagreeable taste, but, by long standing, they begin to fret, and, receiving more acids from the air, recover their former health, and improve in taste.

But should the quantity of oils exceed this last proportion, in wines formed from corn, the must, instead of fermenting, would putrify, even though, by some means, elastic air has been driven into them. In this case, the over proportion of the oil, and its tenacity, prevents the entrance of the acids, the wort receives no enlivening principle from without, and the air, at first conveyed into it, is enveloped with oils so tenacious as to be incapable of action. Nothing so much accelerates putrefaction as heat, moisture, and a stagnating air; and all substances corrupt, sooner or later, in proportion to the inactivity of the contained air, to the want of a proper vent, and to the closeness of their confinement. Besides these cases, beers and ales, as well as wines, sometimes are vapid and flat, without being sour; this does not so much arise from the imbibing the air of the atmosphere, as from their fermenting, generating and casting off too much air of their own. To prevent this accident, they are best preserved in cool cellars, where their active invigorating principles are kept within due bounds, and not suffered to fly off. These facts ought to convince us of the truth, deduced by Dr. Hales, from many experiments, that there is a great plenty of air incorporated in the substance of vegetables, which, by the action of fermentation, is roused into an elastic state, and is as instrumental to produce this act, as it is necessary to the life and being of animals.

I should here close this short and imperfect account; but as, in the art of brewing, there is no part so difficult, and at the same time so important to be in some measure understood, as the cause and effects of fermentation; and as the examination of this act, in all the different lights in which it offers itself to our notice, can hardly be thought uninteresting, these few detached thoughts I hope will be allowed of.

The effect of the act of fermentation on liquors is, so to attenuate the oils; as to cause them to become spiritous, and easily inflammable. When a wine is dispossessed of such oils, which is nearly the case in vinegar, far from possessing a heating or inebriating quality, it refreshes and becomes a remedy against intoxication. The term of fermentation ought, perhaps, only to be applied to that operation which occasions the expressed juices of vegetables to become wine: but as several acts have assumed the same name, it may not be improper here to notice the difference between them.

Vegetation, one of them, is that operation of nature wherein more air is attracted than repelled. I believe all that has been said above, concerning the juice of grapes, is a convincing proof thereof.

Fermentation is, where the communication of the external and internal air of a must is open, and in a perfect state; when the power of repelling, is equal to that of attracting, air.

Putrefaction is when, by the power of strong oils, or otherwise, the communication between the external and the internal air of the must is cut off, so that the liquor neither attracts the one nor repels the other, but, by an intestine motion, the united particles separate and tend to fly off.

Effervescence is when, by the power of attraction, the particles of matter so hastily rush into contact, as to generate a heat which expels the enclosed air; and this more or less in proportion to the motion excited.

SECTION IX.

OF ARTIFICIAL FERMENTATION.

By what has been said, it appears, that, though fermentation is brought on by uniform causes, and productive of similar effects, it is subject to many varieties, both in respect to its circumstances and to its perfection. One difference is obvious, and seems to deserve our attention, as it furnishes a useful division between *natural* and *artificial* fermentation. The first rises spontaneously, and requires nothing to answer all the necessary purposes, but the perfection of the juices, and the advantage of a proper heat. The other, at first sight less perfect, wants the assistance of ferments, or substitutes, without which the act could, either not at all, or very imperfectly, be excited.

There are undoubtedly liquors, which, though they have of themselves a tendency to fermentation, and are naturally brought to it, yet, from some defect in the proportions of their constituent parts, either do not acquire a proper transparency, or cannot maintain themselves in a sound state for a sufficient time. These disadvantages, inbred with them, can hardly ever be entirely removed; they gain very little, especially the latter, from age, and therefore are really inferior to liquors, which require the assistance of substituted ferments, to become real wines. In some artificial fermentations, the ferments are so duly and properly supplied, and so intimately blended with the liquor, that in the end they approach very near to, and even vie with, the most perfect natural wines. Were I to enter into a more minute detail, it might be shewn, that wines, when transported from a hot climate to a cold one, are often hurt and checked in the progress of the repeated frettings they require; from whence they become or remain imperfect, unless racked off from their grosser lees, or precipitated with strong menstruums; whereas beers may be so brewed, as to be adapted either to a hot or a cold region, not only without any disadvantage, but with considerable improvements.

Hitherto I have considered grapes as a most pulpous fruit, sufficient to furnish the quantity of water necessary for extracting its other parts; but the natives of the countries where this fruit abounds, in order to preserve them, as near as possible in their primitive state, after they are gathered, suspend them in barns, or place them in ovens, to dry. Thus, being in great measure divested of their aqueous parts, these grapes remain almost inactive, and without juices sufficient to form wines.

In all bodies, the various proportions of their constituent parts produce different effects; hence they remain more or less in a durable state, and tend either to inaction, fermentation, or putrefaction. Now, by a judicious substitution of such parts as shall be wanting, they are nearly, if not wholly, restored to their pristine nature, as may be proved by the observations and experiments communicated to the public by Dr. Pringle. Thus grapes, though dried and exported from their natural climate to another, by the addition of water only, ferment spontaneously, and form wines very near alike to such as they would have produced before. It may, with confidence, be said, that, when any considerable difference appears, it arises from the injudicious manner in which the water is administered, from the fruit not being duly macerated, or from want of such heat being conveyed to the water and fruit, as the juices would have had, if they had been expressed out of the grapes when just gathered; often from the whimsical mixture of other bodies therewith, and perhaps too from the quantity of brandy, which is always put to wines abroad, to prevent their fretting on board a ship. Upon the whole, though, from what just now has been observed, some small difference must take place, it rather proves than contradicts the fact, that, a due quantity of water being applied to dry raisins, an extract may be formed, which will be impregnated with all the necessary constituent parts the grapes had in them when ripe upon the vine, consequently will spontaneously ferment, and make a vinous liquor. Water then, in this case, becomes a substitute, and the liquors produced in this manner may be accounted of the first class of artificial wines.

Vegetables, in their original state, are divisible into the pulpous and farinaceous kinds, both possessing the same constituent parts, though in different proportions. If from the farinaceous such parts be taken away as they superabound in, and others be added, of which they are defective, these vegetables may, by such means, be brought to resemble, in the proportion of their parts, more especially in their musts, the natural wines I have before been treating of: and these being universally acknowledged to be the standard of wines, the nearer any fermented liquor approaches thereto, by its lightness, transparency, and taste, the greater must its perfection be.

To enquire which of the pulpous or which of the farinaceous kinds of vegetables are fittest for the purpose of wine-making, would here be an unnecessary digression. Experience, the best guide, hath, on the one side, given the preference to the fruit of the vine, and on the other to barley. To make a vinous liquor from barley, having all the properties of that produced from the grape, is a task, which can only be compassed by rendering the wort of these, similar to the must of the other.

As malt liquors require the addition of other substitutes, besides water, to, become perfect wines, they can only be ranked in the *second* class of *artificial* fermentation. These substitutes are properly called *ferments*, and merit the brewer's closest attention.

Ferments, in general, such as yeast, flowers or lees of wine, honey, the expressed juices of ripe fruits, are subjects more or less replete with elastic air, and convey the same to musts, which stand in need thereof. Boerhaave has ranged these, and several others, in different classes, according to their different powers, or rather in proportion to the quantity of air they contain for this purpose.

The juice of the grape, when fermented, forms more lees than the extracts of malt. May we not, from thence, infer that, in the fruit, the elastic air is both more abundant, and contained in a greater number of stronger, though smaller, vesicles, than it is in the malt? The barley, being first saturated with water, germinated only, and then dried with a heat far exceeding that which ripened it, or that which fermentation admits of, has its air in part driven out. The expulsion of air from the worts of beers and ales is still farther effected by the long boiling they undergo. Hence the necessity of replacing the lost elastic air, in order that these extracts may become fermentable. This is effected by means of the yeast, which, consisting of a collection of small bubbles, filled with air, and ready to burst by a sufficient heat, becomes the ferment, which facilitates the change of the wort into a vinous liquor.

The musts of malt generally produce two gallons of yeast from eight bushels of the grain, whereas, in the coldest fermentable weather, and for the speediest purpose, one gallon of yeast is sufficient to work this quantity of malt. Much elastic air still remains in beer, or wine from corn, after the first part of the fermentation is over; for the liquor, separated from the yeast above mentioned, is, at the time of this separation, neither flat, vapid, nor sour; but as yeast, the lees and flowers of malt liquors are of a weaker texture than those of grapes, all artificial fermentations should be carried on in the coolest and slowest manner possible: and beers, but more especially such as are brewed from high-dried, brown malts, (the heat of whose extracts approaches much nearer to that which dried the grain, than is the case in brewing pale malt) ought not to be racked from their lees, as it is frequently practised for natural wines, unless, on account of some defect, they are to be blended with fresh worts under a new fermentation.

As all ferments are liable to be tainted, great care ought to be taken in the choice of them, every imperfection in the ferment being readily communicated to the must. It would not, therefore, be an improper question to be determined by physicians, whether, in a time of sickness, the use of those which have been made in infected places ought to be permitted, and whether, at all times, a drink fermented in a pure and wholesome air is not preferable to that which is made among fogs, smoke, and nauseous stenches.[7]

Wines from corn are distinguished by two appellations, viz., those of ale and beer. As each of these liquors have suffered in character, either from prejudice or want of a sufficient enquiry, it may be proper to levy the objections made against their use, before we enquire into the means of forming them. The most certain sign of the wholesomeness of wines is transparency and lightness; yet some, which are rich, more especially ales, though perfectly fine, have been said to be viscid.—Transparency appears indeed in many wines, before the oils are attenuated to their highest perfection, and some viscidity may therefore be consistent with some degree of brightness. Where the power of the oils and the salts are equal, which is denoted by the transparency of the liquor, viscidity can only arise from the want of age: this cannot be said to be a defect in, but only misapplication of the liquor, by being used too soon.

That beers retain igneous or fiery particles, seems equally a mistake. Malt dried to keep, has undoubtedly its particles removed by fire, so far as the cohesion of them is thereby destroyed, otherwise it would not be in a fit state to preserve itself sound, or readily to be extracted. For this reason, when the grain comes in contact with the water, which is to resolve it, an effervescent heat is generated, which adds to the extracting power, and should be looked on by the brewer as an auxiliary help; but it is impossible that the malt, or the must, should ever inclose and confine the whole or part of fire employed to form them. Fire is of so subtile a nature, that its particles, when contained in a body, continually tend to fly off, and mix with the surrounding air; so that only an equal degree, with what is in the atmosphere, can be continued in the grain, or any liquor whatever, after it has been, for some time, exposed thereto.—Brown beers, made from malt more dried than any other, from experience, are found to be less heating than liquors brewed from pale malt; which probably arises from hence, that brown beers contain a less quantity of elastic air than pale beers, as pale malt liquors contain less than wines, produced from vegetables in their natural state: and as malt liquors contain their elastic air in bubbles of a weaker consistence than those made from the juices of the grape, the effect of beer, when taken in an over-abundant quantity, is neither of so long a continuance, nor so powerful as that of wine, supposing the quality and quantity of each to be equal. This may appear to some persons to be the effect of prejudice, yet it is but a justice due to the produce of my country, to add, that some physicians have given it as their opinion, that strong drinks from malt are less pernicious than those produced from grapes. As far as these gentlemen have, I hope I may advance, without being thought guilty of assuming too much, or countenancing debauch, by pointing out the wines that occasion the fewest disorders.

SECTION X.

OF THE NATURE OF BARLEY.

BARLEY is a spicated, oblong, ventricose seed, pointed at each end, and marked with a longitudinal furrow. The essential constitution of the parts, in all plants, says Dr. Grew, is the same: thus this seed, like those which have lobes, is furnished with radical vessels, which, having a correspondence with the whole body of the corn, are always ready, when moistened, to administer support to the plume of the embryo, usually called the *acrospire*. These radical vessels, at first, receive their nourishment from a great number of glandules dispersed almost every where in the grain, whose pulpous parts strain and refine this food, so as to fit it to enter the capillary tubes; and such an abundant provision is made for the nourishment of the infant plant, that the same author says, these glandules take up more than nine tenths of the seed.

Barley is sown about March, sooner or later, according to the season or soil that is to receive it, and generally housed from ten to twenty weeks after. Most plants, which so hastily perform the office of vegetation, are remarkable for having their vessels proportionably larger; and that these may be thus formed, the seed must contain a greater quantity of tenacious oils, in proportion to those seeds, whose vessels being smaller, require more time to perform their growth and come to maturity. This grain, as may be observed, grows and ripens with the lower degrees of natural heat; from whence, and from the largeness of the size of its absorbent vessels, it must receive a great portion of acid parts. It is said to be viscid, though, at the same time, a great cooler, water boiled with it being often drunk as such; and, however it be prepared, it never heats the body when unfermented.

From these circumstances, of its being viscous and replete with acids, it would at first appear to be a most unfit vegetable, from which vinous liquors, to be long kept, should be made; and, indeed, the extracts from it, in its original state, are not only clammy, but soon become sour.

When the grain is at full maturity, its constituent parts seem to be differently disposed than when in a state of vegetation. By germination alone all its principles are put in action; the fibrous parts possess themselves of a great quantity of tenacious oils, leaving the glandules and finer vessels replete with water, salts, and the purest sulphur. If, in this state, the corn is placed in such a situation, that, by heat, the acid and watery parts may be evaporated, the more such heat is suffered to affect it, the more dry, and less acid, will the corn become; its parts will be divided—its viscidity removed; its taste becomes saccharine, by the acids being sheathed or covered over with oils; and these last be rendered more tenacious in proportion to the greater quantity of heat they are made to endure. This process, regularly carried on, is termed *malting*, and will hereafter be explained more at large.

But, before we enter thereon, it is necessary to consider the state of the grain as it comes from the field.—When mowed, though, upon the whole, it may be said to be ripe, yet every individual part, or every corn, cannot be so. In some seasons, this inequality is so remarkable, as to be distinguished by the eye. The difference in the situation, the soil, and the weather, the changes of the winds, the shelter some parts of the field have had from such winds, are sufficient to account for this, and a much greater variety. When the greater part of the corn is supposed to have come to maturity, it is cut and stacked; the ripest parts having the least moisture, and the fewest acids, as the greenest abound in both. In this state the unripe grains of the corn communicate, to such as are more dry, their moisture and acid parts, which, coming in contact with their oils, an agitation ensues, more or less gentle, in proportion to the power of the acids and water; and from hence is generated a heat, the degree whereof is with difficulty determined.

When this sweating in the mow is kept within its proper limits, the whole heap of the corn, after this internal emotion is over, becomes of one equable dryness, and is not discoloured; but if the grain be put together too wet or too green, the effervescence occasioned thereby will produce such a violent intestine heat, as to charr and blacken the greatest part thereof, nay often make it burst into actual flame.

The effect which a moderate and gentle heat has on the corn, is that of driving the oils towards the external parts of its vessels and skin: by this means, it becomes more capable to preserve itself against the injuries of the weather. The more it is in this state, the backwarder will it be to germinate, when used to this purpose; and if this act is carried too far, or to somewhat like what we have just now mentioned, the plume and root of the enclosed embryo must be scorched, the corn become inert, and incapable of vegetation. This effect is produced by a motion sufficient to remove the particles of the grain from each other beyond their sphere of attraction; and the heat, by which this motion is excited, has been found, in malted corn, to be at about 120 degrees.

It is likely, that vegetables, in general, are susceptible of a large latitude in this respect, according to their different textures. The degree of heat just now mentioned may, perhaps, be applicable to barley alone; the seeds of some grapes endure 126 degrees of heat, and may be capable of being impressed with more, and yet vegetate. But, with corn, if their oils have endured so great a heat, as thereby to be discolored, the seed can by no means be revived. The color of the grain properly indicates the healthy state of the embryo, or future plant; but this, more immediately, is the business of the farmer and maltster, than that of the brewer.

Thus, though it may be disadvantageous to the maltster to steep grain which has not sweated in the mow, as, for want of this, it will not equally imbibe the water; so barley, that is over-heated, or *mow burnt*, cannot be fit for his purpose. It is, in fact, scarcely possible that any large quantity of barley, from the same stack, should make equally perfect malt, as, on its being put together, the heat generated is always greatest in the centre of the rick, and considerably more there than in its exterior parts.

SECTION XI.

OF MALTING.

THIS process is intended to furnish proper means, for setting the constituent principles of the grain in motion: so that the oils, which before served to defend the several parts, may be enabled to take their proper stations.—This is effected by steeping the barley in water, where it strongly attracts moisture, as all dry bodies do; but it requires some time before the grain is fully saturated therewith.[8] Two or three days, more or less, are necessary, in proportion to the heat of the air; for vegetables receive the water only, by its straining through the outward skin, and absorbent vessels, and their pores are so very fine, that they require this element to be reduced almost to a vapor, before it can gain admittance. Heat hath not only the property of expanding these pores, but perhaps also that of adding to the water a power more effectually to insinuate itself.

By the water gaining admittance into the corn, a great quantity of air is expelled from it, as appears from the number of bubbles which arise on its surface when in contact with the grain, though yet much remains therein. A judgment is formed that the corn is fully saturated, so as not to be able to imbibe any more water, from its turgidity and pulpousness, which occasions it readily to give way to an iron rod dropped perpendicularly therein. At this time the water is let to run, or drawn off, the grain taken out of the cistern, and laid in a regular heap, in height about two feet. We have before accounted why moist vegetables, when stacked together, grow hot; so doth this heap of barley. The heat, assisted by the moisture, puts in motion the acids, oils, and elastic air remaining in the corn, and these not only mollify and soften the radical vessels, but, with united power, force the juices from the glandular parts into the roots, which are thereby disposed to expand themselves, and impowered to convey nourishment to the embryo enveloped in the body of the grain. The corn in this heap, or couch, is however not suffered to acquire so great a degree of heat, as to carry on germination too fast, by which not only the finer but also the coarser oils would be raised and entangled together, and the malt when made become bitter and ill tasted; but before the acrospire is perceived to lengthen, the barley is dispersed in beds on the floor of the malt house, and, from being at first spread thin, gradually, as it dries, and as the germination is thereby checked in its progress, it is thrown into larger bodies; so that, at the latter part of this operation, which generally employs two days, much of the moisture is evaporated, its fibres are spread, and the acrospire near coming through the outward skin of the barley. By these signs the malster is satisfied that every part of the barley has been put in motion and separated. It is of great consequence, in making of malt, that the grain be dried by a very slow and gradual heat: for this purpose it is now thrown into a large heap, and there suffered to grow sensibly hot, as it will in about 20 or 30 hours: thus prepared for drying, in this lively and active condition, it is spread on the kiln; where, meeting with a heat superior to that requisite for vegetation, its farther growth is stopped; though, in all probability, from the gentleness of the first fire it ought to be exposed to, none of the finer vessels are, by this sudden change, rent or torn, but, by drying, only the cohesion of its parts removed, rendered inactive, and put in a preservative state. Often, to a fault, the drying of a kiln of malt is performed in 6 or 8 hours: it would be to the advantage of the grain that more than double this time was employed for any intent whatever. It may here be observed, that those oils, which in part form the roots, being with them pushed out from the body of the corn, and dried by heat, are lost to any future wort, not being soluble in water; which is likewise true of those oils which are contained in the shoot or plume; so that the internal part of the malt has remaining in it a greater proportion of salts to the oils than before, consequently are less viscid, more saccharine, and easier to be extracted.

In this process, the acid parts of the grain, though they are the most ponderous, yet being very attractive of water, become weaker, and, by the continued heat of the kiln, are volatilized and evaporated with the aqueous steam of the malt. Thus, by malting, the grain acquires new properties, and these vary at the different stages of dryness; in the first it resembles the fruits ripened by a weaker sun, and in the last those which are the growth of the hottest climates.

When the whiteness of the barley has not been greatly changed by the heat it has been kept in, it is called pale malt, from its having retained its original color; but when the fire in the kiln has been made more vehement, or kept up a longer time, it affects both the oils and the salts of the grain, in proportion to the degree of the heat, and to the time it has been maintained, and thus occasions a considerable alteration in the color. Actual blackness seldom is, and ought never to be, suffered in malts; but in proportion to the intenseness of the fire they have been exposed to, the nearer do they come to that tinge, and from the different brown they shew, receive their several denominations.

The condition the barley was gathered in, whether green or ripe, is also clearly discernible when it is malted. If gathered green, it rather loses than gains in quantity; for the stock of oils in unripe corn being small, the whole is spent in germination, from whence the malt becomes of a smaller body, appears shrivelled, and is often unkindly, or hard. That, on the contrary, which hath come to full maturity, increases by malting, and if properly carried through the process, appears plump, bright, clean, and, on being cracked, readily yields the fine mealy parts, so much desired by the brewer.

The malts, when dried to the pitch intended by the maker, are removed from the kiln into a heap. Their heat gradually diminishes, and, from the known properties of fire, flies off, and disperses itself in the ambient air, sooner or later, as the heap is more or less voluminous; perhaps too in some proportion to the weight of the malt, and as the fire has caused it to be more or less tenacious. Nor can it be supposed that any of its parts are capable of retaining the fire in such a manner as not to suffer it to get away. So subtile an element cannot be confined, much less be kept in a state of inactivity, and imperceptible to our senses. Bars of iron, or brass, even of a considerable size, when heated red hot, cool and lose their fire, though their texture is undoubtedly much closer than that of malt or barley. The experiments made by Dr. Martine, on the heating and cooling of several bodies, leave no room to doubt of this fact, which I should not be so particular about, nor in some measure repeat, was it not to explain the technical phrase used by brewers, when they say, *malts are full of fire, or want fire*. Hence a prejudice hath by some been conceived against drinks made from brown malts, though they have been many months off the kiln, and have no more heat in them, either whole or ground, than the air they are kept in. The truth of the matter is, that, in proportion as malts are dried, their particles are more or less separated from one another, their cohesion is thereby broke, and, coming in contact with another body, such as water, strongly attract from it the uniting particles they want. The more violent this intestine motion is, the greater is the heat just then generated, though not durable. An effect somewhat similar to what happens on malt being united with water, must occur on the grain being masticated; and the impression made on the palate most probably gave rise to the technical expression just taken notice of.

The minute circumstances of the process of malting will be more readily conceived from what will hereafter be said. The effects that fire will have, at several degrees, on what, from having been barley, is now become malt, are more particularly the concern of the brewer; and that these differ, both as to the color and properties, is certain. A determinate degree of heat produces, on every body, a certain alteration, and hence, as the action of fire is stronger or weaker, the effect will not be the same as what it would have been in any other degree.

Barleys, at a medium, may be said to lose, by malting, one fourth part of their weight, including what is separated from them by the roots being skreened off: but this proportion varies, according as they are more or less dried.

As the acrospire, and both the outward and inward skins of the grain are not dissoluble in water, the glandular or mealy substance is certainly very inconsiderable in volume and weight: but as in this alone are contained the fermentable principles of the grain, it deserves our utmost attention.

We have before seen, that wines, beers, and ales, after the first fermentation, are meliorated through age by the more refined and gentle agitations they undergo, and which often are not perceptible to our senses. To secure this favorable effect, we must form worts capable of maintaining themselves, for some time, in a sound state. This quality, however, if not originally in the malt, is not to be expected in the liquor. Some objections have been raised against this method of arguing, and these aided by prejudices, often more powerful than the objections themselves. It is therefore necessary, as malting may be esteemed the foundation of all our future success, to enquire after the best and properest methods of succeeding in this process. Let us, for this purpose, reassume the consideration of the grain, as it comes from the mow, trace it to the kiln, and observe every change it undergoes by the action of the fire, from the time that it receives the first degree of preservation, to that when it is utterly altered and nearly destroyed.

Barley in the mow, though there its utmost heat should not much exceed 100 degrees, may be extracted or brewed without malting. This the distiller's practice daily evinces; but then the extracts, made from this unchanged corn, are immediately put in the still after the first fermentation, else they would not long remain in a sound state. Nor is this method practicable in summer time, as the extracts would turn sour, before they were sufficiently cooled to ferment. It is true, by this means, all the charge of the malt duty is saved; but our spirits thereby are greatly inferior to those of the French.—Boerhaave recommends the practice of that nation, which is to let the wines ferment, subside, and be drawn off fine from the lees, before they are distilled. Was this rule observed in England, distillation would be attempted only from malted grain, which, if properly extracted for this purpose, the difference in the spirit would soon shew how useful and necessary it is to give wines (either from grapes or corn) time to be softened, and to gain some degree of vinosity before they are used to this intent.

But might not barleys be dried without being germinated? Undoubtedly they might; but as they abound with many acids and strong oils, they would require a heat more intense than malt does, before they were sufficiently penetrated, and then the oleaginous parts would become so compact, and so resinous, as nearly to acquire the consistence of a varnish, scarcely to be mollified by the hottest water, and hardly ever to be entirely dissolved by that element.

Barley then ungerminated, either in its natural state or when dried, is not fit for the purpose of making wines; but when, by germination, the coarser oils are expelled, and the mealy parts of the grain become saccharine, might not this suffice, and where is the necessity of the grain being dried by fire? I shall not dwell on the impossibility of stopping germination at a proper period, without the assistance of fire, so that sufficient quantities of the grain, thus prepared, may always be provided for the purposes of brewing; nor even insist upon the difficulty of grinding such grain, as, in this case, it would be spongy and tough. I think it sufficient to mention solely the unfitness of this imperfect malt, for the purpose it is to be applied to, that of forming beers and ales capable of preserving themselves for some time. We should find so many acids blended with the water still remaining in the grain, that, in the most favorable seasons for brewing, they would often render all our endeavors abortive, and, in summer time, make it impracticable to obtain from them sound extracts in any manner whatever.

I have heard of a project of germinating grain, and drying it by the heat of the sun, in summer time, in order, by this means, to save the expence of fuel. Though the hottest days in England may be thought sufficient for this act, as well as for making hay, yet, as barley and grass are not of equal densities, the effects would not be the same. This, however, is not the only objection: as the corn, after a sufficient germination, should be made inactive, this very hot season, favorable, in appearance, to one part of the process, would rather forward, than stop or retard, vegetation; for the barley, by this heat, would shoot and come forward so fast as to entangle too much the constituent principles of the grain with one another, and drive the coarser ill-tasted oils among the finer sweet mealy parts, which alone, in their utmost purity, are the subject required for such as would obtain good drinks.

There often appears in mankind a strange disposition to wish for the gifts of Providence, in a different manner than they have been allotted to us. The various schemes I have just now mentioned, if I mistake not, have sprung from the desire of having beers and ales of the same appearance with white wines. But as they are naturally more yellow or brown, when brewed from malts dried by heats equal or superior to that which constitutes them such, all such projects, by which we endeavour to force some subjects to be of a like color with others, are but so many attempts against nature, and the prosecution of them must commonly be attended with disappointments. It is true, that though the germinated grain be dried slack, yet; if they are speedily used, and brewed in the most proper season, they may make a tolerable drink, which will preserve itself sound for some time: but the proportion, which should be kept between the heat which dried the malt, and that which is to extract it, cannot, in this case, be truly ascertained; and, as the grain will be more replete with air, water, and acids, than it ought to be, the drink, even supposing the most fortunate success, and that it does not soon turn acid, will still be frothy, and therefore greatly wanting in salubrity; for an excess in any of the fermentable principles must always be hurtful.

Barley then, to be made fit for the purpose of brewing, must be malted; that is, it must be made to sprout or germinate with degrees of heat nearly equal to those which the seed should be impressed with when sown in the ground;

and it must be dried with a heat superior to that of vegetation, and capable of checking it. How far germination should be carried on, we have already seen; the law seems to be fixed universally, as to the extent of the acrospire: the degree of dryness admits of a larger latitude, the limits of which shall be the subject of our next enquiry.

Malt dried in so low a degree, as that the vegetative power is not entirely destroyed, on laying together in a heap, will generate a considerable degree of heat, germinate afresh, and send forth its plume or acrospire quite green. The ultimate parts of the nourishing principles are then within each other's power of acting, else this regermination could not take place; and such grain cannot be said to be malted, or in a preservative state. Bodies, whose particles are removed, by heat, beyond their sphere of attraction, can no more germinate; but, coming in contact with other bodies, as malt with water, they effervesce. The grain we are now speaking of first shews this act of effervescence, when it has been thoroughly impressed with a heat of 120 degrees, and a little before its color, from a white, begins to incline to the yellow. Such are the malts, which are cured in a manner to be able to maintain themselves sound, though in this state, and at this degree of dryness, they possess as much air, and as many acid and watery particles, as their present denomination can admit of. This therefore may be termed the first or lowest degree of drying this grain for malt.

To discover the last or greatest degree of heat it is capable of enduring, the circumstance to guide us to it, though equally true, is not so near at hand as effervescence, which helped us to the first. We must therefore have recourse to the observation of that heat, which wholly deprives the grain of its principal virtues. Dr. Shaw observes, *alcohol is one of the most essential parts of wine*; when absent, the wine loses its nature, and, when properly diffused, it is a certain remedy for most diseases incident to wines, and keeps them sound and free from corruption; from whence was derived the method of preserving vegetable and animal substances.—The same excellent author had before this observed, that *no subjects but those of the vegetable kingdom are found to produce this preserving spirit.* Is alcohol, then, a new body, created by fermentation and distillation; or did it originally, though latently, reside in the vegetable? *I have for a good while been satisfied, by experiments,* says Boerhaave, *that all other inflammable bodies are so only as they contain alcohol in them, or, at least, something that, on account of its fineness, is exceedingly like it; the grosser parts thereof, that are left behind, after a separation of this subtile one, being no longer combustible.*

Now, as the same author has clearly proved[9] that fire, by burning combustible bodies, as well as by distilling them, separates their different inflammable principles, according to their various degrees of subtilty, the alcohol residing in the barley, when exposed to such a degree of heat as would cause it to boil, i. e. 175 degrees, must make great efforts to disengage itself from the grain. Is it not, therefore, natural to conclude, that, in a body like malt, whose parts have been made to recede from one another, (from whence it is porous, and easily affected by fire,) prepared for fermentation, or the making a vinous liquor, this event will probably happen at the same time when the body of the grain has been ultimately divided by fire, or that malt charrs? and if this is true, may not charring be termed the last degree of dryness, when, even somewhat before it takes place, the acid parts and finest oils, which are necessary for forming a fermentable must, fly off, and cannot be recovered.—Charring seems to be a crisis in solid bodies, somewhat analogous to ebullition in fluids; both being thereby perfectly saturated with fire, their volatile and spiritous parts tend to fly off. In charring, the subject being ultimately divided by fire, the constituent principles are set at liberty, and escape in the atmosphere, in proportion to their several degrees of subtilty, and to the fire which urged them. In boiling they are equally divided, and incline to disperse; but, even the more volatile, being surrounded with water, a medium much denser than themselves, they are caught up therein, and, by the violent motion caused in boiling, entangled with it, and with other parts it contains, so as not to be extricated or divided therefrom except by the act of fermentation. Now, as liquors boil with a greater or less fire in proportion to their tenacity and gravity, solid bodies may likewise be charred by various proportions of heat. The whole body of the barley, as its different parts are of different texture, cannot, at the same instant, become black, nor, where any quantity of the grain is under similar circumstances, if not equally germinated, can the whole charr with the same degree.

To the several reflections, before made, I thought proper to add the surer help of experience. I therefore made the following trial, with all the care I was capable of. If the effects of it appear satisfactory, by gaining two limited and distant degrees, we may determine and fix the properties of the intermediate spaces, in proportion to their expansion.

In an earthen pan, of about two feet diameter, and three inches deep, I put as much of the palest malt, unequally grown, as filled it on a level to the brim. This I placed over a little charcoal, lighted in a small stove, and kept continually stirring it from bottom to top.

At first it did not feel so damp as it did about half an hour after. In about an hour more, it began to look of a bright orange color on the outside, and appeared more swelled than before. Every one is sensible that a long-continued custom makes us sufficient judges of colors, and this sense in a brewer is sufficiently exercised. Then I masticated some of the grain, and found them to be nearly such as are termed brown malts. On stirring, and making a heap of them, towards the middle, I placed therein, at about half depth, the bulb of my thermometer; it rose to 140 degrees: the malt felt very damp, and had but little smell.

At 165 degrees, I examined it in the same manner as before, and could perceive no damp; the malt was very brown, and on being chewed, some few black specks appeared.

Many corns, nearest the bottom, were now become black, and burnt; I placed my thermometer nearly there, and it rose to 175 degrees: but, as the particles of fire, ascending from the stove, act on the thermometer, in proportion to the distance of the situation it is placed in, through the whole experiment an abatement of five degrees should be allowed, as near as I could estimate.— Putting, a little after, my thermometer in the same position, where about half the corns were black, it shewed 180 degrees. I now judged that the water was nearly evaporated, and observed the heap grew black apace.

Again, in the centre of the heap, raised in the middle of the pan, I found the thermometer at 180 degrees; the corn tasted burnt, the surface appeared, about one half part a full brown, and the rest black. On being masticated, still some white specks appeared, which I observed to proceed from those barley-corns which had not been thoroughly germinated, and whose parts cohering more closely together, the fire, at this degree, had not penetrated. The thermometer was now more various, as it was nearer to, or farther from, the bottom; and, in my opinion, all the true-made malt was charred, for their taste was insipid, they were brittle, and their skins parting from the kernel.

I, nevertheless, continued the experiment, and, at 190 degrees, still found some white specks on chewing the grain; the acrospire always appearing of a deeper black, or brown, than the outward skin; the corn, at this juncture, fried at the bottom of the pan.

I still increased the fire; and the thermometer, placed in the middle, between the bottom of the pan, and the upper edge of the corn, shewed 210 degrees. The malt hissed, fried, and smoked abundantly. Though, during the whole process, the grain had been kept stirring, yet, on examination, the whole was not equally affected by the fire. A great part thereof was reduced to perfect cinders, easily crumbling to dust between the fingers, some of a very black hue, without gloss, some very black, with oil shining on the outside. Upon the whole, two thirds of the corn were perfectly black, and the rest of a deep brown, but more or less so, as the grains were hard, steely, or imperfectly germinated. This was easily discovered by the length of the shoot: most of the grains seemed to have lost their cohesion, and had a taste resembling that of high-roasted coffee.

In the last stage of charring the malt, I placed over it a wine glass inverted, into which arose a pinguious oily matter, and tasted very salt. It may, perhaps, not be unnecessary to say, that the length of time this experiment took up, was four hours, and that the effect it had, both on myself, and on the person who attended me, was such as greatly resembled that of inebriation.

Though, from this experiment, the degree of heat at which malt charrs, is not fixed with the utmost precision, yet we see that black specks appeared, when the thermometer was at 165 degrees; some of the corns were entirely black at 175, others at 180. In proportion as fire causes a deficiency of color, it must occasion a want of fermentable properties, the whole of which are certainly dispersed, when the grain becomes of an absolute black. Thus we may conclude, with an exactness surely sufficient for the purposes of brewing, that true germinated malts are charred in heats, at about 175 degrees: as these correspond to the heat at which pure alcohol, or the finest spirit of the grain itself, boils, it seems to require this heat, wholly to extricate itself from the more tenacious parts of the corn; which, when deprived of this etherial enlivening principle, remains inert, incapable of forming a fermentable must or wort, and indicates to us, that the constituent parts of vegetables may be resolved by heats, equal to those between the first degree which formed them, and the last, which ultimately destroys their properties; though the extracts will possess different qualities or virtues, according to the determinate heat which is applied.

SECTION XII.

OF THE DIFFERENT PROPERTIES OF MALT, AND OF THE NUMBER OF ITS FERMENTABLE PARTS.

THE consequences resulting from the before-mentioned experiment have already been hinted at. But it is necessary to trace them farther, and to shew how much they tend to the information and use of the brewer.

Germinated barleys, so little dried, as that their particles remain within their sphere of attraction, are not in a preservative state, and therefore cannot properly be termed malts.

The first degree of dryness, which constitutes them such, as we have seen before, is that which occasions them to cause some effervescence. This cannot be effected, when they are dried with less than 120 degrees of heat; the highest that leaves them white. When urged by a fire of 175 degrees, they are charred, black, and totally void of fermentable principles. Now this difference of heat, being 55 degrees, and producing in the grain so great an alteration, as from white to black, the different shades or colors, belonging to the intermediate degrees, cannot, with a little practice, be easily mistaken.

White, we know, from Sir Isaac Newton's experiments, is a composition of all colors, as black is owing to the absence of them. These two terms indicate the extremes of the dryness of malt. The color, which the medium heat impresses upon it, is brown, which, being compounded of yellow and red, the four tinges which shade malt differently, may be said to be white, yellow, red and black. The following table, constructed on these principles, will, on chewing the grain, readily inform the practitioner of the degree to which his malts have been dried. It is true some doubts have arisen, whether the increase of heat is by equal divisions (according to the scales marked on thermometers) or whether the degrees should not rather be in proportional parts: but if the effect of fire on bodies (as every experiment shews) is exactly corresponding to the expansion it is the cause of, this undetermined question in no wise affects the brewery.

A TABLE *of the different Degrees of the Dryness of Malt, with the Changes of Color occasioned by each Increase of the Degrees.*

Degrees.		
119	White	White
124	W. W. Yellow	White turning to a light Yellow.

129	W. W. Y. Y.	Yellow.
134	W. W. Y. Y. Red,	High yellow.
138	W. W. Y. Y. R. R.	Amber.
143	W. Y. Y. R. R.	Light brown.
148	Y. Y. R. R.	Brown.
152	Y. R. R.	High brown.
157	Y. R. R. Black,	Brown inclining to black.
162	Y. R. R. B. B.	High brown, speckled with black.
167	R. R. B. B.	Half brown, half black.
171	R. B. B.	Coffee color.
176	Black,	Black.

N. B. The several letters against each degree, it is apprehended, will help in practice to fix the color.

The foregoing table not only enables us to judge of the dryness of the malt by its color, but also, when a grist is composed of several sorts of malt, to foresee the effect of the whole when blended together by extraction. Some small error may possibly occur in judgments thus formed upon the report of our senses; but as malt occupies different volumes, in proportion to its dryness, if, in the practice of brewing, upon mixing the water with the malt, the expected degree is observed, such parcel of malt may be said to have been judged of rightly, in regard to its dryness. So that the first trial either confirms or corrects our opinion thereof.

Though malt, dried to 120 degrees, is in a preservative state, yet is it the least so as malt: it then possesses the whole of its fermentable principles, which, if not impeded in the extraction, would be very speedy and active: the duration of the worts to be formed from grain so low dried, must entirely depend on the power given to the water by heat, to draw from the malt, oils of such consistence as shall sheath and retard the hasty effects of the fermentable parts. By extraction, then, malted grain, even so low dried as this, may, with very hot waters, and with the farther assistance of hops, be made to produce beers, which for years will be capable of maintaining themselves sound, or for a long time to resist the effects of the hottest climates. They may also, by a less heat being given to the extracting water, and blended with less hops, form drinks, which shall be fit for use in so short a time as a week, and perhaps a term much shorter: hence we see the degree of heat which dried the malt, and the degree of heat given to the water to extract it. The mean of these numbers (making an allowance for the quantity of hops used) is that which directs us to fix the properties and duration of the wort. In one sense, then, we may consider malt, so low dried as this, as being such as would in the shortest time furnish us with a fermented liquor, and in another, such as would yield the most delicate and strongest drink. When malt charrs, and becomes black, its parts are ultimately divided; it has lost the principles fit to form a fermentable wort, and which it once possessed. The degree of heat, prior to that which produces this effect, is the last which still retains any part of the fermentable properties. In worts from malt thus highly impressed by fire, fermentation would proceed with so slow and reluctant a pace, that, in this case, they might be said to be in the utmost state of preservation. No term can be fixed for their duration. A liquor of this sort, brewed with the greatest heat it would admit of, in the extracting water, might keep many years, and become rather accommodated to the temperature of the place it was deposited in, than to its own constituent parts. Experience has shewn, that drinks, impressed by the drying and extracting heat, with a medium of 148 degrees, with a proper addition of hops, at the end of eighteen months, have been found sound, and in a drinkable state; and at this degree we find the middling brown.—From these two extremes, and on these principles, the following table is formed, exhibiting the length of time drinks made from malt, impressed with each respective degree of heat, properly brewed, in the most favourable season, will require, before they come to their due perfection to be used.

Equally as with hot extracting waters, low dried pale malt may be made to yield beers which will long continue in a sound state; so high dried malt, acted upon by cooler and low extracting water, may be made to furnish a wort soon fit for use, though less agreeable and more inelegant. It might here be asked, why, then, at any time, is malt dried with heats exceeding 120 degrees? In answer to this, it might justly be said, it would be very difficult for the malster exactly to hit this point of drying, without deviating from it either on the one side or on the other; and suppose this difficulty removed, still he could not be certain every individual grain was equally affected: if the drying was less than 120 degrees, the malt, by receiving the moist impressions of the air, would regerminate, and be spoiled. Before the use of hops, malt was high dried, as a means to keep the extracts sound. To eradicate an ancient custom or prejudice requires a long time. This, and the conveniency of keeping malts, was the reason why, for many years, it was in general dried to excess; an error which for some time past has been losing ground, as no reason at present subsists, why malts should exceed in color a light amber.

A TABLE, *shewing the age beers will require, before being used, when brewed from malts, which, in drying and extracting, have been impress'd with a medium heat corresponding to the following degrees.*

Degrees.	Shortest time with 12 lb. of hops.		Longest time with 12 lb. of hops.				Shortest time wit the fewest quantity of hops possible.	
119	2	Weeks						
124	1	Month	3	Months	}	Brewed in the proper saison	{	2 Weeks
129	3	Months	6	Months				4 Weeks
134	4	Months	9	Months				6 Weeks
138	¹⁰6	Months	12	Months				6 Weeks
143	7	Month	3	Months	}	Brewed in summer	{	2 Weeks
148	9	Month	3	Months				2 Weeks
152	10	Months	18	Months				
157	18	Months	2	Years				
162	2	Years						
167								
171								
176								

It must be observed, that the foregoing table is constructed on the supposition, that these different sorts of malt are brewed, fermented with the utmost care, with waters heated to extract it, in proportion to the dryness of the grain, and to intent of time there set down, and with an adequate addition of hops; an ingredient which shall be considered in its proper place. What is meant by the water being heated to extract malt in proportion to the dryness of the grain, may merit some explanation.

Grapes, when ripe, carry with them the water they have received, both during their growing state, and that of their maturity. This quantity is sufficient to form their musts with. To dried grapes or raisins, water is added, to supply what they have lost; and for the same reason it is requisite in regard to malt: but as grapes stand in no need of artificial fire, to give to their fermentative principles a due proportion, so what they produce themselves, or cold water applied to them, when dry, is a sufficient menstruum. But barleys, wanting the assistance of a great heat to bring their parts to the necessary proportion, require, when malt, a similar or rather a greater heat to resolve them: without which, experiment shews, the flour of the grain would come away undissolved, and thus considerably impoverish the grist.—Should, on the other hand, too great a heat be applied, an equal loss would be sustained, from some of the finer parts being coagulated or blended with oils, tenacious beyond the power of fermentation to exhibit them. The proportioning therefore the heat of the water to the dryness of the malt, more especially to obtain from the grain the whole strength it is capable of yielding, as well as to cause the drink to preserve itself sound its intended time, is of real necessity.

Well-brewed drinks should not only preserve themselves sound their due space, in order to be meliorated by time; they should likewise be fine and transparent.—These circumstances prove the artist's skill and care, as well as the salubrity of the drink; and are the surest signs of a well-formed must, and of a perfect fermentation. If then the rules for obtaining these ends can be deduced from the foregoing principles and experiments, we may flatter ourselves with possessing a theory, which will answer our expectations in practice.

According to the laws of nature discovered by Sir Isaac Newton, the spaces between the parts of opaque bodies are filled with mediums of different densities, and the discontinuity of parts, each in themselves transparent, is the principal cause of their opacity. Salts in powder, or infused in an improper medium, will intercept the light; gums make a muddy compound, when joined to spirits; and oils, unassisted by salts, refuse to be incorporated with water. Musts, therefore, whose constituent parts are not capable of being dissolved by water into one homogeneous body, are not fit, either for a perfect fermentation, or a pellucid drink.

Length of time, which improves beers and wines, often rectifies our errors in this respect; for the oils being, by various frettings, more attenuated, and more intimately mixed, the liquor is frequently restored, and becomes of itself pellucid. Yet I never found this to succeed, where the error upon the whole of the dryness of the malt, and the heat of the extracts, exceeded the medium by 10 degrees.

Art has also, in some measure, concurred with nature to remedy this defect. When beers or wines have been suffered to stand, till they are rather in an attracting than in a repelling state, that is, when their fermentations and frettings apparently stand still; then, if they do not become spontaneously fine, they may be precipitated, by mixing with them a more ponderous fluid.

The floating particles, that occasioned the foulness, are, by this means, made to subside to the bottom, and leave a limpid wine: but the power of dissolved isinglass, the ingredient generally used for this purpose, seldom takes effect, when the error exceeds the medium, as before, by more than 10 degrees.

Other ingredients, indeed, have been used, which carry this power near 10 degrees farther. It is not my province to determine, whether such be salutary: undoubtedly it would be better if there were no occasion for them. Beyond these limits, precipitation has no effect; the liquor, which cannot be fined thereby, if attempted, by increasing the quantity of the precipitants, will be overpowered by the menstruum, and injured in its taste. How frequent this last case of cloudiness is, would answer no purpose in this place to enquire. The use of doubtful ingredients, and such errors as have been mentioned, need no longer blemish the art, when a constant and happy practice, will be both the effect and the proof of a solid and experimental theory.

Beers which become bright of themselves, or by time alone, as well as those precipitated either by dissolved isinglass, or by more powerful means, each possess their respective properties in a certain latitude or number of degrees; and as these effects arise wholly from the heats employed in drying the malts, and in forming the extracts, the following table will be of use to point out the limits, within which each drink may be obtained.

A TABLE, *shewing the tendency beers have to become fine, when the malt, in drying and extracting, has been impressed with heats, the medium of which answers to the following degrees, supposed to be brewed and kept in the most eligible manner.*

Deg.				
119	White,	} Immediately.	}	Latitude of musts which fine spontaneously.
124	Inclining to yellow,			
129	Yellow,	2 Months.		
134	High yellow,	4 Months.		
138	Amber,	6 Months.	}	Latitude of musts which fine by precipitation.
143	Light brown,	8 Months.		
148	Brown,	10 Months.		
152	High brown,	12 Months.		
157	{ Brown, inclining to black,	} 14 Months.	}	Latitude of heats which cannot

162	{	High brown speeckled with black	}	18 Months.	form musts, so as to answer the intent of becoming wholesome beer.
167		High brown half black		18 Months.	
171		Coffee color,	}	20 Months.	
176		Black,			

The difference between the heat for forming grapes, and the greatest heat which ripened them, affords to us the number of degrees answerable to their constituent parts: the investigation of barley, in like manner, though less important to our purpose, yet may, with some propriety, be admitted.

Upon examination it will be found, barley ears, and the new grain begins to form (being still in possession of its flower) about the same time with us as grapes do, in June; when we found the mean heat of the air in the shade to be 57.60 degrees.

Barleys in general are mowed from August to September; so that, in their growth, they are benefited by the whole of our summer's heat, and for like reasons as in page 59, we estimate this 61.10 degrees: 3.50 degrees then would be the number of their constituent parts, taken from the degrees of heat in the shade, and which perhaps would be different if the actual sun-shine heat and what is reflected from the earth, were accounted for. Barleys are annuals, unbenefited by the whole of the autumn sun; but, after being mowed, they are stacked, retaining still much of their straw, leaves, and outward skins. In these heaps they heat, more or less, according to the condition in which they were housed; and which heat may reach to 120 degrees or more, but in general is equal, or somewhat superior, to that of our bodies. The properties of the grain, by this means improved, ripen, and from hence are more capable of preserving themselves. This might be a reason why a farther allowance should be made to the number of degrees denoting their constituent parts: how much, by a very great number of observations, made from the germination, ripening, to the stacking of the barley, in many years, and in many cases, might probably be ascertained; but the difficulty of doing this, and afterwards the impossibility of complying with the information such enquiries would afford, and the little need there is for it, as nature has allowed a considerable latitude for our deviating from what may be styled perfection, without any sensible injury: these circumstances render such enquiries unnecessary, if not fruitless.

Vegetables, but more particularly barley, from their first origin to such time as they might be ultimately separated by fire, may be divided into different periods, according to the distinct properties belonging to each, (and each of these require again a more exact enquiry.) Barley is under the act of germination, so long as the acrospire or stem is within the outward skin of the parent corn; this excluded, it vegetates so long as it receives nourishment by the interposition of its roots. It may be said to be in a state of concentration, when receiving but little or no support from the earth, yet it is acted upon by such heats as do not exceed what it might bear in the vegetative period; and in that of inaction, when, by the power of heat, it is placed in a passive state. Now malt is barley germinated, and, by a quick transition, is impressed with heats superior to those admitted in vegetation, and such as places the corn in a state of inaction. In the beginning of the process of malting, the more tenacious oils, together with some salts, are excluded from the body of the grain, to form the vessels requisite to forward the growth of the future plant. What remains in the parent grain (that choice food, at first necessary to the infant barley) are saccharine salts, alone applicable to the brewer's purpose, and of the nature and quantity of which, he ought to be well acquainted. To retain these, and prevent a waste thereof, the germinated corn is placed in such heat, as destroys the union between its parts, from whence it becomes inactive. When this intent is obtained by the least heat capable of effecting it, the malt retains both its color, and the whole of its properties.

Vegetables, in no part of their growth, are ever affected by heats so great as to disperse their constituent parts; on the contrary, by natural heats, in general they are improved. The whole of their elements then, must be measured from the first degrees which form them, to the last which procure their highest perfection; and in climates where they are not benefited by the whole of such heat, their properties must be accounted only so many degrees, as in such places are between the extremes of their germination and maturation. Alike with malt, their whole number of constituent parts, denoted by degrees of heat, must be so many as are comprehended between that degree which leaves it in possession of the whole of their elements, and the first heat which excludes a part; for malt more dried than this, being less perfect, and losing some of its properties, fewer must remain.

The degree of heat which in malt divides the period of germination from that of inaction, we have found to be 119; the grain then is perfectly white, and shews little if any sign of effervescence; the first change, fire occasions therein, is to impress it with a light yellow color; this takes place at 129 degrees of heat, an alteration which can proceed from no other cause, but, in removing its original whiteness, to have expelled some of its primitive parts. The difference then between these two numbers of 10, specifies, in degrees of Fahrenheit's scale, the number of properties constituting barley, malt.

It must be confessed this is establishing a principle of the art of brewing, upon the uncertain report of our senses, as perhaps our sight may deceive us in fixing this change of color exactly at 129 degrees; but we know white and black to be the two extremes of the dryness of malt, and that the middle color between them is brown, which being compounded of yellow and red, these four tinges, equally divided, as we have done in the foregoing tables, will corroborate our fixing the teint of yellow at this degree. The table shewing the tendency beers have to become fine, was formed from experiments made on brewings, whose governing medium heats were from 134 to 148, the proportion in point of time given by these, justifies the division between immediate pellucidity, at 119, and that taking place at two months, or 129 degrees. So from hence we may be satisfied, however an absolute perfection cannot be depended upon, yet this being the most exact division our senses afford, it approaches so near to truth, that if any mistake remains, it can be but trivial, compared to the latitude of errors, fermentation and time correct. But this number, 10 degrees, denoting the quantity of fermentable parts, must lessen in proportion as a continued, or a greater heat deprives the grain of more properties. A speedy spontaneous pellucidity is the effect of the whole fermentable parts; malt affected by heat, conveyed either through air or water, or through both, (so the medium of these exceeds not 138 degrees,) if assisted by the acids gained to the drink by long standing, such will obtain transparency. Beers, then, intended to be formed of themselves to become fine, in the calculations used to discover their elements, so many of the members of the constituent parts must be implied, as corresponds with the time the beer is intended to be kept; but when beers are made intentionally to require precipitation to become fine, in such proportion as we purpose to impress opacity on the drink, we must, in the calculations made to discover the temperature of the extracts, imply only so many of the constituent parts, as correspond to the medium heat which will occasion this foulness. These few observations shew the necessity of establishing this fundamental doctrine, the use of which will obviously appear in practice.

Thus does the success of this art depend on the instrument so often mentioned, which, by indicating the expansions caused by different heats, becomes a sure guide in our operations. I shall now close this account, by comparing with the principles here laid down, the defects which we, but too often, meet in barley when malted.

SECTION XIII.

OBSERVATIONS ON DEFECTIVE MALTS.

IN the preceding enquiry, some of the defects of malt have been occasionally mentioned: but as a perfect knowledge of the grain, especially when it has undergone this process, is a matter of no small concern to the brewer, I shall now bring such defects into distinct view, both to compare them with the foregoing principles, and that the knowledge of them may be more at hand, on every occasion, when wanted.

Every different degree of heat acting on bodies causes a different effect: and this varies also, as such heat is more or less hastily applied. The growth of vegetables is in general submitted to these laws: but yet I conceive there is some difference between germination and vegetation, which I beg leave to point out. The former seems to be the act caused by heat and moisture, while the plume or acrospire is still enveloped within the teguments of the parent corn, and it is most perfectly performed by the gentlest action, and consequently by the least heat, that is capable of moving the different principles in their due order. Vegetation, again, is that act which takes place when the plant issues forth, and, being rendered stronger by the impressions of the air, becomes capable of resisting its inclemencies, or the warmth of the sun-shine. Germination is the only act necessary for malting, the intention being solely to put in motion the principles of the grain, and not to rear up the embryo to a plant. Now, as this begins in barley at the degree where the water first becomes fluid, or nearly so, the cold season, when the thermometer shews from about 32 to 40 degrees, would seem the most proper for this purpose. How far its latitude may with propriety be extended, experience alone can determine. Maltsters continue to work so long as they think the season permits, and leave off generally in May, when the heat of the water extends at a medium from 50 to 55 degrees. But the nearer they come to this medium, with the greater disadvantage must they malt: as, by such warmth, the vessels of the corn are much distended, the motion of the fluids violent, and the finer parts too apt to fly off. Thus the coarser oils gaining admittance, the glandular parts become filled with an impure and less delicate sulphur, which, instead of a sweet, inclines to a bitter, taste. This is so manifest, and so universally experienced, that, in general, brewers carefully avoid purchasing what is termed *latter-made malts*.

Malt, which has not had a sufficient time to shoot, so that its plume may have reached to the extent of the inward skin of the barley, remains overburthened with too large a quantity of earth and oils, which otherwise would have been expended in the acrospire and radical vessels. All those parts of the corn which have not been separated, and put in motion by the act of germination, will, when laid on the kiln to dry, harden and glutinize: no greater part thereof will be soluble in water, than so far as the stem or spire of the barley rises to, or very little farther, and as much as is wanting thereof will be lost to the strength of the drink.

When malt is suffered to grow too much, or until the spire is shot through the skin of the barley, which is not often the case, though all that is left be malt, that is, containing salts dissoluble in water, yet as too large a portion of oils has been expended out of the grain, such malts cannot be fit to brew drinks for long keeping.—There is, besides, a real loss of the substance of the corn, occasioned by its being overgrown.

Malt, the germination of which has reached and been stopped at the proper period, and has been duly worked upon the floors, if not sufficiently dried on the kiln, even though the fire be excited to a proper heat, retains many watery parts. The corn, when laid together, will be apt to germinate afresh, perhaps to heat so as to take fire; should not this extreme be the case, at least it must grow mouldy, and communicate an ill flavor to the drink.

Malt, well grown, and worked as before, but over-dried, though with a proper degree of heat, will become of so tenacious a nature, as to require a long time before it can admit of the outward impressions of the air to relax or mellow it, that is, before it is fit to be brewed with all the advantages it otherwise would have; and in proportion as it has black specks on being masticated, so much of its parts being charred is a diminution to the strength of the liquor, besides impressing it with a burnt or nauseous taste.

Malt, dried on a kiln not sufficiently heated, must require proportionably a longer time to receive the proper effect of the fire; the want of which will bring it into the same state as malt not thoroughly dried.

If too quick or fierce a fire be employed, instead of gently evaporating the watery parts of the corn, it torrifies the outward skin, divides it from the body of the grain, and so rarifies the inclosed air as to burst the vessels. Such is called *blown malt*, and, by the internal expansion, occupies a larger space than it ought. If the fire be continued, it causes its constituent parts to harden to the consistence of a varnish, or changes it into a brittle substance, from whence the malt is said to be steely and glassy: it dissolves but in a small proportion, is very troublesome and dangerous in brewing, and frequently occasions a total want of extraction; by the brewery termed, *setting the grist*.

Malt, just, or but lately, taken from the kiln, remains warm for a considerable time. Until the heap becomes equally cool with the surrounding air, it cannot be said to be mellow, or in a fit state to be brewed: its parts being harsh and brittle, the whole of its substance cannot be resolved, and the proper heat of the water, which should be applied to it for that purpose, is therefore more difficult to be ascertained.

The practice of those maltsters, who sprinkle water on malt newly removed from the kiln, to make it appear as having been made a long space of time, or, as they say, to *plump* it, is a deceit which cannot too much be exposed. By this practice, the circumstance of the heat, and harshness of the malt, is only externally, and in appearance, removed, and the purchaser grossly imposed on. The grain, by being moistened, occupies a greater volume, and, if not speedily used, soon grows mouldy, heats, and is greatly damaged.

The direct contrary is the case of malt which has been made a long time: the dampness of the air has relaxed it, and so much moisture has insinuated itself into the grain, that some doubt must arise how much hotter the mash should, for this reason, be. Yet, supposing no distemper, such as being mouldy, heated, or damaged by vermin, it is observed, malt, under this circumstance, may more certainly be helped in brewing, than those just abovementioned.

From what has been said, it appears how necessary it is to procure malt which has been properly steeped, germinated to its true pitch, and dried by a gentle, moderate heat, so as the moisture of the corn be duly evaporated, then cured by just so much fire as to enable it to preserve itself a due time, without being blown or burnt. How easy it is to regulate this process in the cistern, in the couch, on the floors, and on the kiln, when the malster, intends no artifice to save his excise, I need not say; but with what certainty and ease the whole might be carried on by the help of the thermometer, I leave such to determine as are modest enough to think, that the art may be brought to more accurate rules than those of the bare report of our unassisted senses. As such rules may easily be deduced from the principles here laid down, I shall not be more particular in shewing their application, as not being my immediate purpose, nor my business as a brewer: nor have I leisure, or the conveniency of a malt house, to make experiments of this sort; yet with truth it may be said, that such as would not be disappointed in their brewing, must take care not to be deceived in their malt. This, however, being but too frequently the case, we should constantly be on our guard against its defects, and know how to correct them. If it is treated in the same manner as if it was perfect, the well-malted parts alone will be digested. If too slack dried, it may be corrected by an addition of heat, if over-dried, or injured by fire, it may proportionably be helped. By applying the thermometer to the extracts, more particularly to the first, the brewer thereby will be informed, to a sufficient degree of exactness, of the defects he can mend, and hardly be ever at a loss for the properest means to work the grain to the greatest advantage.

As far as we have proceeded in our enquiry, though some satisfaction must arise from our being enabled to account for the greater part of the process of brewing, yet it may be observed, even with the assistance of the thermometer, as yet a geometrical exactness, in many respects, has not been attained; but nature, when the interest and necessities of mankind are the object, apparently has supplied our wants, and rectified our defects. In this art, fermentation, when allowed to display itself, corrects all our errors to a considerable latitude, though as yet, of this act, it may be said we scarcely conceive its cause, or properly discern its effects.

PART II.

THE PRACTICE OF BREWING.

Before I enter upon the practical, and indeed most important, part of this work, it will not be improper to give a distinct, though general, view of the different parts it is to consist of.

To extract from malt a liquor, which, by the help of fermentation, may acquire the properties of wine, is the general object of the brewer, and the rules of that art are the subject of these sheets.

An art truly very simple, if, according to vulgar opinion, it consisted in nothing else than applying warm-water to malt, mashing these together, multiplying the taps at discretion, boiling the extracts with a few hops, suffering the worts to cool, adding yeast to make it ferment, and trusting to time, cellars, and nostrums, for its taste, brightness, and preservation!

A few notes and observations, such as are too often found to be foisted under the articles of beer and brewing, in some books of agriculture and others of cookery, might be sufficient, were the place and constitution of the air always the same, the materials and vessels employed entirely similar, and lastly, the malt drinks intended for the same use and time; but, as every one of these particulars is liable to variations, and can be complied with, only by the application of different determinate heats; was the artist to submit himself to loose, vague, and erroneous directions, like those above mentioned, they would only serve to deceive him, and his case would be but little mended, if he trusted to indefinite signs, and insufficient maxims, in his deviation from them.

A more certain foundation has been laid down in the first part of this treatise, and the principles there established will, I trust, in all cases, answer our ends, provided we make use of proper means to settle their application. The most elegible means to effect this, must be to follow, as near as possible, such plan, which the rational brewer would, in every particular circumstance, sketch to himself, before he proceeded to business. His first attention ought to be directed not only to the actual heat of the weather, but also to that which may be expected in the season of the year he is in. The grinding of his malt must be his next object, and as the difference of the drinks greatly depends upon that of the extracts, he cannot but chuse to have distinct ideas of what may be expected from the amount of the heat of them. Hops, which are added as a preservative to the extracts form too important a part to be employed without a sufficient knowledge of their power. The strength of malt liquors depending principally on their quantity or lengths, it is necessary to ascertain the heights in the copper, to answer what, on this account, is intended. The difference in boiling, for different drinks or seasons; the loss of water by evaporation; the proper division of the whole quantity of this element employed, and, in proportion to such division, that of the heat to be given in each part of the process; the means to ascertain these degrees, by determining what quantity of cold water is to be added to that, which is at the point of ebullition, come afterwards under his consideration. The manner and time of mashing, the many expected incidents which must produce some small variations between the actual and the calculated heat of his extracts, it will be incumbent upon him to make a proper estimation and allowance for. To dispose of the worts in such forms and at such depths, as may render the influence of the ambient air the easiest and most efficacious, and then, by the addition of yeast, to provide the drink with that internal and most powerful agent it had lost in boiling, are the next requisites. Fermentation, which follows, and which the brewer retards or forwards according to his intentions, completes the whole process; after these necessary precautions, to compare his operations with those of the most approved practitioners in his art, and to find himself able to account for those signs and established customs, which before were loosely described, authoritatively dictated, and never sufficiently determined or explained, must be to him an additional satisfaction. As precipitation is requisite in certain cases, the common methods for effecting it should be known, and likewise the means practised among coopers to correct the real or imagined errors of the brewer, in order to render the drink agreeable to the palate of the consumers, will naturally lead him to consider what true taste is, and by employing the means, by which it may safely be obtained and improved, he will have done all in his power, to answer his customers expectation, and to secure his success.

This arrangement, which appears the most simple, is that, which the reader will find observed in the following sections. The proper illustrations of tables and examples have not been omitted, and from the complete plans for brewing, under two forms of the most dissimilar kind, it will be found the rules are adapted to all circumstances, and applicable to every purpose.

I must here add somewhat in justification, for publishing what may be said to be the mysteries of an art, often too cautiously precluded from the sight and attention of the public; but every art and science whatever have equally been laid open, and from such communication received greater improvements, and become more useful to mankind in general, and the professors of them in particular. If attention is given to the rules and practice here laid down, it will be found that the brewer, from the large quantities he manufactures, from repeated experience, from the conveniency of his utensils, and more than all, from the interest he has to be well acquainted with his business, is most likely to be successful, in preference to any one else, and therefore can have no reason to be displeased on being presented with a theory and practice, which, far from being the sole right of the brewery, the discovery of the principles were certainly the property of the author and of his friends, whose names would do his work honor if mentioned. From the application of these principles, being convinced of their exactness and facility in practice, he offers his labor to a trade he esteems, with no other view than the hope he entertains of being of some service to it and to the public.

If, notwithstanding repeated endeavours, some things, in this treatise, should appear out of their places; others, in more than one; if redundancies, chiefly occasioned by the natural temptation of accounting for particular appearances, have not always been avoided; if inaccuracies should now and then have escaped me, let it be remembered (by the good-natured it certainly will) that, in new and intricate subjects, digressions and repetitions are in some measure allowable, that an over-fulness is preferable to an affected and often obscure brevity, and that the improvement of the art, rather than the talent of writing, must be the brewer's merit, and was my only aim.

SECTION I.

OF THE HEAT OF THE AIR,

AS IT RELATES TO THE PRACTICAL PART OF BREWING.

IN and about the city of London, the most intense cold that has been observed is 14 degrees, and the greatest heat has made the thermometer rise, in the shade, to 89. Within these limits are comprehended all the fermentable degrees, and consequently those necessary for carrying on the process of brewing. If the lowest degree proper for fermentation be 40, and the highest 80, the medium of these two would, at first sight, appear to be the fittest for this purpose; but the internal motion, necessary to carry on fermentation, excites a heat superior to the original state of the must by 10 degrees. Hence, if 60 degrees be the highest eligible heat a fermenting must should arise to, 50 should be the highest for a wort to be let down at, to begin this act; which heat can only be obtained, when that of the air is equal thereto, so that it denotes the highest natural heat for beers and ales to be properly fermented. With regard to the other extreme, or the lowest heat, however cold the air may be, as the worts, which form both beers and ales, gain, by boiling, a degree greatly superior to any allowed of in fermentation, it is constantly in the artist's power to adapt his worts to a proper state. The brewing season, then, may justly be esteemed all that part of the year in which the medium heat of the day is at or below 50 degrees: this, in our climate, is from the beginning of October to the middle of May, or 32 weeks; the most elegible period of time for brewing all kinds of beers.

But, as many incidents often make it necessary to extend these limits, the only time for venturing to comply therewith is, when the medium heat of the season is at 55 degrees; by which, six weeks more may be obtained. But, under these circumstances, the quantity of beer brewed should be less, that the worts may cool more readily, by being thinner spread; and, to gain more time, the brewing is best carried on with two worts only: taking these precautions, and beginning early in the morning, the first wort, by laying long enough in the coolers, will, towards evening, be brought to a heat of 55 degrees. The night, in this season of the year, being generally colder by 10 or 12 degrees than the medium heat of the whole 24 hours, the second worts may be reduced to a cold of 43 degrees: the mean of 55 and 43, being 49 degrees, would be the real heat of the worts in the ton; and with 10 degrees more, (the heat gained by fermentation,) still it would not reach 60 degrees, the highest fermentable heat, beers intended to preserve themselves long should arrive to; but so near would it be to this, and so little is the uniformity of the heat of the air to be relied on, that necessity alone can justify the practice of brewing such drinks, when the heat of the air is so high as 55, consequently, where it exceeds this, it should never be attempted.

As the extractions are made by heats far superior to any natural ones, though the actual temperature of the air neither adds to, nor diminishes from, their strength, yet it is to be known for the following reason. The proper heat given to the mash is by means of cold added to boiling water; and cold water generally is of no other heat than that of the air itself. Indeed, when the cold is so intense, as to occasion a frost, and to change water into ice, that which is then used for brewing, being mostly drawn from deep wells, or places where frost never, or but seldom, takes place, may be estimated at 35 degrees, and this will be sufficiently exact.

The following table shews the temperature of the air for every season in the year, and confirms what I have just now said concerning the season proper for brewing, and the actual heat of the water. It was deduced from many years' observations, made with very accurate instruments, at eight o'clock in the morning, the time in which the heat is supposed to be the medium of that of the whole day.

A TABLE, *shewing the medium heat, for every Season of the year, in and about London, deduced from observations made from 1753 to 1765, at eight o'clock each morning.*

Degrees.				Degrees.			
January to	1 15	}	36' 38	July to	1 15	}	60' 52
to	31	}	34' 97	to	31	}	34' 97
February to	1 14	}	35' 51	August to	1 15	}	59' 89
to	28	}	38' 11	to	31	}	38' 48
March to	1 14	}	37' 99	September to	1 15	}	55' 17
to	28	}	39' 72	to	31	}	54' 13
April to	1 14	}	43' 13	October to	1 15	}	48' 66

to	28	}	46´ 04		to	31	}	46´ 72
May	1	}	49´ 05		November	1	}	42´ 26
to	14				to	15		
to	28	}	55´ 67		to	31	}	39´ 40
June	1	}	57´ 20		December	1	}	38´ 61
to	14				to	15		
to	28	}	59´ 14		to	31	}	37´ 54

To ascertain the authority of this table, and to make it useful to several purposes, I have carried to decimals the mean numbers resulting from my observations.—But such an exactness has been found, in the practice of brewing, to be more troublesome than necessary. I have therefore constructed another table, similar to the former, but where the fractions are omitted, and the whole numbers carried on from five to five. The heats of the latter end of October, and beginning of November, have here been set down rather higher than they really are; as, at this time of the year, the hops fit to brew with are old and weak, and I could not devise any means more easy to allow for their want of strength.

A TABLE, *shewing the medium heat of the air, in and about London, for every season of the year, applicable to practice.*

Degrees.					Degrees.			
January	1	}	35		July	1	}	60
to	15				to	15		
to	31	}	35		to	31	}	60
February	1	}	35		August	1	}	60
to	14				to	15		
to	28	}	40		to	31	}	60
March	1	}	40		September	1	}	55
to	15				to	15		

to	31	}	40		to	30	}	55
April	1	}	45		October	1	}	50
to	15				to	15		
to	30	}	45		to	31	}	50
May	1	}	50		November	1	}	45
to	14				to	15		
to	31	}	60		to	31	}	40
June	1	}	60		December	1	}	35
to	14				to	15		
to	28	}	60		to	31	}	35

As nothing is so inconstant as the heat of the air, we are not to be surprised when it deviates from the progression specified in the table. The flowing water used in the brewery, at the coldest seasons, we have fixed at 35 degrees, and the highest heat in the air, to carry on the process for beers brewed for long keeping, at 55 degrees. The length proper to be drawn, or the quantity of beer to be made from each quarter of malt being fixed, the brewer, at any time, has it in his power to make calculations for brewings, supposing the mean heat of the air to be at 35, at 40, at 45, at 50 and even at any degree of heat whatever, so as never to be unprovided for any season. Water, being a body more dense than air, requires some time to receive the impressions either of heat or cold, for which reason the medium heat of the shade of the preceding day, will most conveniently govern this part of the process, unless some very extraordinary change should happen in the atmosphere. This must make the business of the artist, in this respect, very easy, as, in the course of his practice, he will have only to correct the little changes that occasional incidents give rise to; and the calculations will answer all his purposes so long as the lengths of beer to be brewed from the same quantity of malt remain unaltered, and with very little variation and trouble, when the coppers employed, by being changed, are of different dimensions.

The best method to know the true heat of cold water, would be to keep a very accurate and distinct thermometer, in the liquor back; but as this, in every place, is not to be expected, and inaccuracies must arise from a change in the air, to prevent their consequences in practice, we must have recourse to experience. This has taught us that a difference of 8 degrees, between the actual heat of the water, and that from which the brewing was computed, will produce, in the first extract, a difference of four degrees.

Most brewers' coppers, though they vary in their dimensions, are generally made in proportions nearly uniform; the effect of one inch of cold water more or less, will therefore nearly answer alike, that is, it will alter the heat of the tap, by 4 degrees. But this will only hold good in such cases, where the water is in the same proportion to the volume of the grist. In brewing brown beers, or porter, three worts are generally made; the extracts therefore must be of different lengths from what they are in beers brewed at two worts only. In this case, the quantity of water for the first wort, is less than it otherwise would be; and what must be allowed for the first mash, to wet the malt, is so much as to occasion the second, or piece liquor, to be proportionably less also; as it is of great consequence, if the first tap doth not answer to its proper degree, that the second should be brought to such a heat, as to make up the medium of the first and second extracts, the second, or piece liquor, by reason of its shortness, is more conveniently, and more exactly tempered in the little copper; and one inch cooling in, is in this case found, both by calculation and experience, to occasion a difference of one degree of heat only in the mash.

One of the principal attentions, in forming beers and ales of any sort whatever, is that they may come to their most perfect state, at the time they are intended to be used. Common small beer is required to be in order, from one to four weeks, and as it is impossible to prejudge the accidental variations, as to heat and cold, that may happen in any one season of the year, it is rational to act up to what a long experience has shown, is to be expected, and to mix such quantity of cold water with that, which is made to come to ebullition, as to bring the extract to the degree fixed for each particular season, let the heat, at the time of brewing, vary therefrom, in any degree whatever.

In treating on the subject of air, in the former part of this work, I observed the effect it had in penetrating the parts of the malt, or in the technical term used by brewers, in slacking it. As such is the case, when the grain is entire and whole, it is more so when ground, and experience teaches us, that, when malt has been about 24 hours from the mill, the dampness it has imbibed is equal to half an inch more of cold water added to that which is to be made to boil for the first liquor, and produces therefore a diminution of 4 degrees in the heat of the tap[11].

An effect, somewhat resembling this, is caused by the impression of the air on the utensils of a brewhouse, which are not daily used; the heat received from a foregoing process has expanded their pores, and rendered them more susceptible of cold and moisture. From this circumstance, the heat of the first mash will be affected in a proportion equal to half an inch less cooling in, or in the space of 24 hours, to 4 degrees of heat.

The time of the day, in which the first extract is made, becomes another consideration; for as 8 o'clock in the morning is the time of the medium heat in the whole 24 hours, the other hours will give different degrees. When a first mash is made about 4 o'clock in the morning, the following table shews the difference between the heat at 4 and 8; that of the other hours, in the like case, may be learned by observation. It has been observed, that, in the cold months, from the sun's power being less, the heat of the day and night are more uniform, and also that the coldest part of the 24 hours is about half an hour, or an hour before sun-rising. I have judged it convenient to place, in the same table, the several incidents affecting the first extract.

INCIDENTS *occasioned by the air affecting the heat of the first extract, to be noticed more particularly, when small beer is brewed, as the quantity of water is then greatest, and the mash more susceptible of its impressions*

Morning at 4 o'clock*		
January	0	Utensils, for want of being used, in 24 hours lose 4 degrees of heat, equal to half an inch of cold water.
February	0	
March.	2	
April	4	Malt, which has been ground 24 hours, imbibes moisture equivalent to half an inch, which lessens the heat by 4 degrees.
May	6	
June	8	
July	10	The difference between the actual heat of the air, and that naturally expected is to be allowed in proportion of 8 degrees to one inch cooling in.
August	8	
September	6	
October	4	

November	2	Malts, from having been long kept, or old, become considerably slacked.
December	0	

* Colder by so many degrees than at eight o'clock in the morning.

Before we quit this subject, it may not be improper to observe, that, in the hottest season, and in the hottest part of the day, the difference between the heat of the air in the shade, and that in the sun's beams in and about London, is nearly 16 degrees, and also that cellars or repositories for beers, are, in winter, generally hotter by ten degrees, than the external air; and in summer, colder, by five.

SECTION II.

OF GRINDING.

MALT must be ground, in order to facilitate the action of the water on the grain, which otherwise would be obstructed by the outward skins. Every corn should be cut, but not reduced to a flour or meal, for, in this state, the grist would not be easily penetrable. It is therefore sufficient that every grain be divided into two or three parts, nor is there any necessity for varying this, for one sort of drink more than another. In every brewing the intention of grinding is the same; and the transparency of the liquor, mentioned by some on this occasion, depends, by no means, on the cut of the corn.

It has been a question, whether the motion of the mill did not communicate some heat to the malt; should this be the case, it can be but in a very small degree; and, what may arise from hence, will be lost by shooting the grain out of the sacks, or uncasing the grist into the mash ton. Of late years it has been recommended, instead of grinding the malt, to bruise it between two iron cylinders: if, by this means, some of the fine mealy parts are prevented from being lost in air, it must be very inconsiderable, and, perhaps, not equal to the disadvantage of the water not coming in immediate contact with the flour of the grain. In brewing, not all, but only a certain portion of the constituent parts of the malt are requisite; these, heated water alone is sufficient to procure, so that, upon the whole, the difference between bruising and grinding the grain can be of no great consequence.

We have before observed, malt, by being ground and exposed for some time to the air, more readily imbibes moisture than when whole, and the dampness, thus absorbed, being in reality so much cold water, a grist, that has been long ground, is capable of being impressed with hotter waters than otherwise it would require. In country places, where the quantity brewed consists only of a few bushels of malt, and make so small a volume as to be incapable to maintain an uniform heat, where the people are ignorant, that a certain degree is necessary to form a proper extract with; and where, instead of this, boiling water is indifferently applied, the effects of these errors are in some measure prevented, by grinding the malts a considerable time, as a month or six weeks before the brewing, and by the excess of fire readily escaping from so small a quantity. This method, from the inconstant state of the air, and from the impossibility of acting up to rule, must be very uncertain and fortuitous, so that few or no arguments are necessary to explode it. The truth is, the merit of country ales, so often mentioned, proceeds from the forbearing to use the drink, but when it is in the fittest state. Thus time not only corrects the errors of the operators, but also gives them, in the eyes of the consumers, the credit of an extraordinary knowledge and unmerited ability.

SECTION III.

OF EXTRACTION.

FIRE impressed on malt, either through air or water, it is true, has similar effects as to preservation, but the fact is not the same as to taste: the sweet, the burnt flavor, or the proportion of both, the malt originally had, sensibly appear in the extracts; but water heated to excess will not, in extracting pale malt, communicate to the worts an empyreumatic taste; whether this proceeds from some acid parts, still residing in the heated waters, which might help the attenuated oils to tend towards a sweet, or from other reasons, is not easily determinable; certain it is, the foundation of taste in malt liquors is in the malt itself.

The basis of all wines is a sweet: this circumstance for brewing beers agreeable to the palate must always be attended to. Next to this, it is required that the liquor should possess all the strength, it can fittingly be made susceptible of. Pale malt, as it retains the whole virtue of the grain, yields the strongest beers. The finest oils being fittest for fermentation, malt dried by fierce heats, in a great measure loses these, and what remains are not only coarser oils, less miscible with water, but such as bring with them the impressed taste of fire.

To answer the purposes of taste, strength, and preservation, from what has been said it appears, that the extracting water must be of a heat superior to that which dried the malt; no other rule appears to direct in this, than to make choice of malt of such dryness, the delicacy of which has not been removed by fire, and such as will, at the same time, admit of a sufficient number of superior degrees of heat, to extract all its fermentable parts; that is (see page 124) malt whose dryness is nearly 19 degrees less than the mean of the drying and extracting heats applicable to the purpose intended.

As 119 degrees, the first heat forming pale malt, and at which it possesses the whole of its sweetness and virtues, may be said to be the lowest degree of dryness in the grain to form keeping beers with, so 138 degrees, above which the native whiteness of the grain is so subdued, as to remain but in a very small proportion, is the highest dried malt fit to be used for any purpose; from these premises the following table is formed, to shew the degree of dryness of malt, where taste and strength are equally consulted, to brew drinks capable of keeping themselves sound a long time, at any medium required.

The proper choice of malt I thought necessary to point out, previous to entering more at large on the subject of extraction. This table, it must be observed, is in no wise directive for brewing common small beer, soon to be expended, that liquor depending on many other circumstances, of which notice will be taken immediately under that head.

A TABLE, *shewing the proper dryness of Malt, applicable to the mean of the drying and extracting heats under which keeping malt liquors should be formed.*

Mean degrees of dryness of malt and heat of extracts.	Color of malt expressed in degrees.
138	119
140	121
142	123
144	125
146	127
148	130
150	132
152	134
155	136
157	138

The subject to be resolved having been examined as to its dryness, we now come to the immediate matter for which this section was intended.

Extraction is a solution of part, or the whole, of a body, made by means of a menstruum. In brewing, it is chiefly the mealy substance of the grain that is required to be resolved; fire and water combined are sufficient to perform this act. Water properly is the receptacle of the parts dissolved, and fire the power, which conveys into it a greater or less proportion of them.

When all the parts necessary to form a vinous liquor are not employed, or when more than are required for this purpose are extracted, the liquors must vary in their constituent parts, and consequently be different in their effects. This difference arises either from heat alone, or from the manner of applying it; and the properties of beers and ales will admit of as many varieties as may be supposed in the quantity of the heat, and in its application. But as the useful differences are alone necessary to the brewer, they may be reduced to the four following modes of extraction.

First, that which is most perfect, and for which malt is chose of such dryness, in which it with certainty possesses the whole of its constituent parts, and the extracts are made with such heats, as to give the beer an opportunity to be improved by time, and to become of itself fine and transparent.

Secondly, that from which, in order to obtain every advantage of time, strength, and flavor, such extracts are produced as cannot become pellucid of themselves, but require precipitation.

Thirdly, that which is intended soon to become intense, where soundness and transparency are for some short time expected, but not always obtained, because brewed in every season of the year, and deprived of the advantages which age and better managements procure to the first.

Fourthly, that where the advantages of strength and pellucidity are to be procured in a very short space.

These four modes of resolving the grain, being the fundamental elements on which almost every specie of drink is brewed, I must observe, the two first may be said to be an exact imitation of natural wines, in forming which, the principles we have laid down may fully be applied. The third is the effect of necessity, by which we are deprived of that time nature directs for properly producing fermented liquors, and where we are subjected to many disadvantageous circumstances; to guard against the consequences of which, we must rely, in some measure, upon opinion formed from observation alone; and the fourth may be said to be art too precipitately carried on. Before I treat of them separately, it is requisite to mention a few general rules applicable to all.

In the enquiry we made of the means which nature employs to form the juices of grapes, we found two remarkable circumstances: the first, a necessary lesser heat for the production of the fruit, and the second, a much greater for its maturation; the former useful to incline the must to fermentation, the latter to raise therein such oils as should maintain it for some time in a sound state. But in all wines, an evenness of taste is requisite to affect the palate with an elegant sensation; and it may be observed, the autumn and spring heats being nearly equal; so the first juices of grapes are formed by almost, uniform impressions; the summer heats, though stronger, act upon the same principle; for though the grapes remain upon the vine some part of the autumn, perhaps in this space they gain little more than the juices prepared by the summer's sun: from whence the tastes of wine are more simple than otherwise they would be. Thus are we directed, that a first wort shall have the least share of heat of the whole brewing, and the last wort the greatest; intermediate worts; if any; must be proportioned to both, and if several mashes of extracts are made to compose a wort, these must be equal as to their heat, being careful at the same time to preserve to the process the medium heat which is to govern the whole. By this means, we shall obtain our intended purpose; and place into the drink one and the same smooth taste.

In the table[12] shewing the different effects produced in the grain by the different degrees of heat, the numbers, with respect to beers, express, not only the mean of the degrees of dryness the malt had, with those also of heat in the extracting liquors, but also is implied the power communicated by the hops, that is, it imparts to us, the idea of the whole combination.

As malt liquors are made with different views, so must the principles on which they are formed be varied. Beers intended long to be kept, require more heat in their extracts, in order to produce such oils, or so many in quantity from the grain, as shall retard and delay the quick effects of fermentation; and malt liquors, which are soon to be brought into use, claim an opposite management. This is imitating nature, for we have before observed[13], the hotter the autumnal, the vernal and maturating heats are, with more power do the wines resist the impressions of time and the air; and we traced the rule which governed this variety, by an enquiry into the number of degrees required to form the juices of grapes, and applied their number to discover the first and last heats they were impressed with. In calculations to find out the heat to be given to water properly to resolve the malt, the same method must be followed, it being equally necessary here to employ only such a proportion of the number of degrees which constitute the whole of the fermentable principles in malt that are needful to the purpose we would answer. We have said malts continue in possession of all their constituent parts from their first degree of dryness, 119 to 129. By age alone beers obtain spontaneous pellucidity, when urged in the whole of their process with a heat so great as 138 degrees, precipitation or art extends it to near 157 degrees, after which neither the acid parts furnished by the air, nor art avails: an obstinate foulness is the result; from whence it may be concluded, that at or beyond this heat, so great a part of the fermentable principles is dispersed, as what remains in the grain has not power sufficient to produce transparency. The following table, founded on these principles, will hereafter be found directive to fix the first and last heats to be given to the extracts of malt.

A TABLE, *shewing the quantity of fermentable principles residing in malts at their several degrees of dryness, or, the number of constituent parts which form beers in proportion to their properties*[14], *specified in degrees, and to be used in calculations, made to ascertain the proper heats to be given to the first and last extracts of malt.*

Mean degrees of heat affecting malt.	Constituent parts.
119	10
124	9
129	8
134	7
138	7
143	6
148	5
152	5
157	4
162	3
167	2
171	1
175	0

Though beers and ales are divided into strong and small, this division regards only the proportion of the vehicle, and not that of the constituent parts. The same means, as to the heat of the extracts, must be employed, to form small beers, capable of preserving themselves sound for some time, as are used to make strong drinks: for though a small liquor possesses more aqueous parts, the oils and salts of the malt are only more diluted, not altered in their proportions, and this causes but a very small difference in the duration of the liquor.

It now remains to apply these rules, deduced from the theory, to the several sorts of malt liquors, which answer to the four modes of extraction, just before laid down.

The first and most perfect is, when the malt is chosen of such dryness, and the extracts made with such heats, as give the beers an opportunity of being improved by time, and slow fermentations, to become spontaneously bright and transparent. Under this head, may be comprehended all *pale keeping strong*, and all *pale keeping small* beers.

From its name, regard must be had to the color of the malt, and such only used, as is dried the least, or by 119[15] degrees of heat.

The hops should likewise be pale, and their quantity used in proportion to the time the drink is intended to be kept; suppose, in this case, it is 10 months, 10lb. of fine hops, for every quarter of malt, will be required.

The highest degree of heat, or rather the medium of the highest dryness in malt, with the mean heat of the several extractions, to admit of spontaneous pellucidity, we have seen in the foregoing table (page 124) to be 138 degrees, and this medium is chosen, as it answers not only the intent of long keeping, but of brightness also.

From the medium degree of the malt's dryness, and of the heat of the extracts, to determine the heat of the first and the last extract, and the value in degrees of the quantity of hops to be used, for brewing pale strong and pale small beers, intended to be kept about ten months before they are used, and expected to become self-transparent.

119	Malt's dryness.
——	
138	Mean of malt's dryness, heat of extracts, and value of hops.
3	Degrees, value of 10 lb. of hops.
——	
135	Mean of malt's dryness and heat of extracts.
For the first liquor.	
135	As before.
3½	Half the number of the constituent degrees, answerable to 138 degrees, the mean heat of the whole process, to be subtracted[16].
——	
131½	Degrees governing the first extracts.
——	
119	Malt's dryness.
144	First rule to discover the heat of the first extract.
263	
——	
131½	As above.
——	

	For the last liquor.
135	As before.
3½	Half the number of the constituent degrees, to be added, to find
——	
138½	The degrees governing the last extract.
——	
119	Malt's dryness.
158	First rule to discover the heat of last mash.
——	
277	
——	
138½	As above.
——	

The elements for forming pale strong and pale small beers, intended to be kept, are therefore as follows:

Malt's dryness.	Value of hops.	Whole medium.	First heat.	Last heat.	
119	3	138	144	158	
			2	2	heat lost at the time the extract separates from the grist.

The proof of this is as follows:

144	Heat of the first extract.
158	Heat of last extract.
——	
302	
——	
151	Mean heat of extracts.

119	Malt's dryness.
——	
270	
——	
135	Mean heat of Malt's dryness, and of heat of extracts.
3	Value of hops.
——	
138	Whole mean given as above.
——	

It is necessary to add 2 degrees to the heat of every mash, such being the mean of 4 degrees, constantly lost in every extract, at the time they are separated from the grist, and exposed to the impressions of the air.

The second mode of extraction is, that, in which every advantage which can be procured from the corn, from art, and from time is expected; this produces such drinks, as cannot become spontaneously pellucid, but require the help of precipitation.

The improvement, which every fermented liquor gains by long standing, is very considerable; the parts of the grain, which give spirit to the wine, being, by repeated fermentations, constantly attenuated, not only become more light and pungent, but more wholesome. If, in order to give to beers more of the preservative quality, greater quantities of oils are extracted, in proportion to the salts, transparency cannot take place; but, when the heat employed for this purpose does not exceed certain limits, this defect may easily be remedied, and the drink be fined by precipitation; as time enables it to take up part of the very oils, which at first prevented its transparency, it will, by long standing, and by precipitation, become both brighter and stronger.

Where the demand for a liquor is constant and considerable, but the quantity required not absolutely certain, it ought to be brewed in such manner that time may increase its merit, and precipitation render it almost immediately ready for use. These circumstances distinguish this class of extraction, and justify the preference given to *porter* or *brown* beer, which comes under the mode we are now treating of.

Though transparency in beers is a sure sign of the salts and oils being in an exact proportion, it is in no wise a proof of the justness of taste: for strong salts acting on strong oils may produce pellucidity, but the delicacy and pungency of taste, depend on the finer oils and the choicest salts being wholly preserved, these best admitting of fermentation, and most perfectly becoming miscible with the liquor, the more volatile oils and salts of the grain if excluded, by the malt being too high dried, the consequence in the beer must be, an heavy and rancid taste. The less dried the malts are, which are brewed for beers to be long kept, the hotter are the extracts required to be, but this greater heat being communicated to the grain through water, an element eight hundred times more dense than air, the finer parts of the corn, though acted upon by an heat which in air would disperse them, by this means are retained.

It appears, by the table (page 124) that drinks brewed from malts, affected by heats, whose medium is 148 degrees, and with twelve pounds of hops to every quarter of malt, require from 6 to 12 months with precipitation to become bright; this is the age generally appointed for brown beers to be drank at, and by the table, page 133, we find the proper malts where the medium heat of the whole process is 148 degrees, must be such as have been dried with 130 degrees to form this liquor, whose color as yet is expected to be full or brown, without being deprived of more valuable qualifications.

In the drink before examined, the number of degrees which constitute the properties of malt, affected by a mean heat of 138 or 7 degrees, were employed, they being intended to become, in time, spontaneously bright; but, as this quality in the present case is required only with the assistance of precipitation, the number 5, in the table, shewing the constituent parts remaining in the grain at every degree of dryness, (page 168) as this corresponds to the medium 148, is undoubtedly that which must answer our purpose, both as to the nature and to the time this liquor is in general made use of. These conditions being premised, the proper degrees of the first and last extract for porter will be found by the same rules as were used before.

130	Degrees, malt's dryness.
———	
148	Degrees, whole medium intended.
———	
4	Degrees, value of hops, fractions omitted.
———	
144	Mean of malt's dryness and heat of extracts.

	For the first extract.
144	As before.
2½	Half the number of the constituent degrees to be deducted.
141½	Mean of malt's dryness, and of the heat of the first extract.
130	Malt's Dryness.
153	Rule to discover the first heat.
283	
141½	As above.
	For the last extract.
144	As before.
2½	Half the number of the constituent degrees to be added.
146½	Mean of malt's dryness, and of the heat of the last extract.
130	Malt's dryness.
163	Rule to discover the last heat.
293	
146½	As above.

The elements for brewing brown strong beers, with two degrees added to the first and last extracts, for what is lost at their parting from the malt, independent of its farther division into the respective mashes.

Malt's dryness.	Value of hops.	Medium heat of the extracts, malt's dryness, and value of hops.	First heat.	Last heat.
130	4	148	155	165

Brown beers, brewed with malt so low dried as 130 degrees, twenty years since, would have appeared very extraordinary, and most likely, at that time, when a heaviness and blackness in the drink formed its principal merit, would have been a sufficient reason to condemn the practice; but strength and elegance being now more attended to, have justified the brewer, in making porter, to employ malt of such degree of dryness, as he shall think will best answer these purposes.

As high liquors used to extract low dried malt will form a must capable to preserve itself equally a long time, as an adequate liquor used to high dried malt doth; and the first of these methods having greatly the advantage of the other in point of taste, as 130 degrees of dryness in malt is one, from its change of color, where part of its finer principles may be supposed to be evaporated. It may not be amiss to enquire if there be not reasons why malt, less affected by fire, should be used for manufacturing this commodity.

The medium of the malt's dryness, and of the heat of the extracts, together with the value of the hops which are to make porter, is 148 degrees. This, because precipitation has been found convenient and necessary for this drink, yet, when at the proper age, it has undergone this last operation, it is supposed to shew itself in its best form; bright, well-tasted, and strong; that is, in such state as drink should be, which becomes spontaneously transparent, and is capable of preserving itself a long time, if from

148 degrees.

The value of the oils yielded by the hops (See page 180) is deducted,   4 degrees.

———

Will remain,   144

And by table (page 162) we find a must under the mean of 144 degrees should be formed with malt dried to 125 degrees, with this circumstance the elements of brewing porter will be as follows.

125	Malt's dryness.
———	
148	Degrees, whole medium intended.
4	Value of hops.
———	
144	Mean of malt's dryness, and heat of extracts.
	For the first extract.
144	As before.
———	
2½	Half the number of constituent parts, to be deducted.
———	
141½	Mean of malt's dryness, and of the heat of the first extract.
———	
125	Malt's dryness.
158	Rule to discover the first heat.
———	
283	
———	
141½	As above.
	For the last extract.
144	As before.
2½	Half the number of constituent parts, to be added.
———	
146½	Mean of malt's dryness, and of the heat of the last extract.
———	
125	Malt's dryness.
168	Rule to discover the last heat.
———	

293	
———	
146½	As above.

Elements for brewing porter with malt dried to 125 degrees, and two degrees added to the first and to the last extracts, for what heat is lost at their parting from the malt, but this, independent of a farther allotment of this heat to the respective mashes.

Malt's dryness.	Value of hops.	Medium of the heat of the extracts, malt's dryness, and value of hops.	First mash.	Last mash.
125	4	148	160	170

Whether any attempt to improve this liquor, by using malt of less dryness than 125 degrees, may ever be put in practice, is very uncertain; porter, if brewed with malts so low as 119 degrees, probably would succeed; for, in this case, the last mash, according to the foregoing rules, would be at the 174th degree, at which the spirit of the grain could not be dispersed, and probably the result would be, a more delicate, more strong, and more vinous liquor.

It may be observed, that 4 degrees are charged for the quantity of hops used; as this number corresponds to the quantity proper to form beer of this denomination. A greater or a less proportion of hops is sometimes allowed to this drink, on account of its better, or inferior quality, of the necessity there may be to render it fit for use in a shorter time than that which is commonly allowed—from nine to twelve months, and, lastly, of old, stale, or otherwise defective drinks, blended, with new guiles. In these cases, which cannot be too rare, the errors should be corrected only by the addition of hops, and no alteration be made, either in the dryness of the malts, or in the heat of the extracts.

The third mode of extraction is that which intends spontaneous transparency, but not a durable liquor. Under this head is comprehended *common small beer*, soon to be drank.

Common small beer is supposed to be ready for use, in winter, from two to six weeks, and in the heat of summer, from one week to three. Its strength is regulated by the different prices of malt and of hops; its chief intent is to quench thirst, and its most essential properties are, that in the winter it should be fine, and in the summer sound. This liquor is chiefly used in and about great trading cities, such as London, where, for want of a sufficient quantity of cellar room, drinks cannot be stowed, which, by long and slow fermentations, would come to a greater degree of perfection. The duration of this kind of liquor being short, and there being a necessity of brewing it in every season of the year, dividing it into very small quantities, easily affected in its conveyance by the external heat: generally neglected, and placed in repositories influenced by every change of air, the incidents attending it, and the methods for carrying on the process must be more uncertain, various, and complicated, than those of any other liquor made from malt.

The incidents attending this specie of malt liquor are so many, so short of existence, so contrary to one another, and often so different from what should be expected in the different periods of the year, that an attempt to guard, in a just proportion, against every one of them, and against what *may* happen, and oftentimes does not, must be fruitless. After many endeavours of this sort, which terminated in a doubtful success, we have found it most eligible to form these drinks in proportion to the principal circumstances constantly attending them, and the result was more fortunate, as, in general, the drink was able to maintain itself against that variety of temperature it met with in the places allotted to it.

In proportion as it is brewed, in a hot or in a cold season, we must employ every means, either to repel or to attract the acids circulating in the air; for this purpose, the degree of dryness in the malt, the quantity of hops, the heat of the extracts, and the degree of temperature the wort is suffered to ferment with, must vary as such seasons do. The success, in brewing common small beer, greatly depends on its fermentation being retarded or accelerated, in proportion to the heat of the air, and expansion being the principal effect of heat, was a wort of this sort suffered, in winter, to be so cold as 40 degrees, the air would, with difficulty, if at all, penetrate the must, or put it in action. This slow fermentation would not permit the beer to be ready at the time required.— For these reasons, brewers let down their worts, in that season, at 60 degrees, whereas, in summer, the air of the night is made use of to get them as cold as possible, by which means a part of them may be 12 degrees colder than the medium of the heat of the day, and the whole of the worts nearly 5 degrees, in the space of 24 hours.

The choice of the malt, as to its dryness and color, for brewing this liquor, should be varied in proportion to the several seasons, but custom requires it should be kept nearly to an uniform color. For this reason, when the air is so cold as the lowest fermentable degree, a greater dryness than 119 degrees is required; but the dryness of malt forming only one part of the process, the proper medium directing the whole must be brought to its true degree, by the heat given to the extracts. In the height of summer, malt dried to 130 degrees seems to be the best, as it unites the properties of speedy readiness, preservation, and transparency, and these several characters are, at that time, requisite in this liquor.

To come as near as possible to the inclination of the consumers, or to maintain as near as may be an uniform color, if in the hottest season malt dried to 130 is best for this purpose, the mean between this and 119, the first degree that constitutes malt, must answer nearest every intent, when the heat of the air is at 40 degrees. Upon this footing, the following table will, from the proportion of these two extremes, shew the color of the grain for every season of the year.

Heat in the air. Malt's dryness. Value of hops in degrees.

Heat in the air	Malt's dryness	Value of hops in degrees
35	122	1
40	124	1
45	125	1
50	127	1
55	129	1½
60	130	2

If common small beer was immediately to be used after being brewed and fermented, and it was free from the incidents, most of which we have just now enumerated, no hops would be required, and the medium degree of the whole process would be that of the lowest dried malt, 119, to be employed when the heat of the air was at its first fermentable degree, or 40, as, with adequate malts, this would make the liquor that would be ready in the least space, and, at the same time, yield its constituent parts; but if small beer was intended to be kept some short time, brewed without hops, and not liable to any accidents, and the process to be carried through, in a heat of air equal to the highest fermentable degree, or 80, in this case the governing medium for the whole process must be the utmost heat the grain is able to endure, where malt charrs, or 175 degrees. As malt liquors are principally affected by heat, we will first proportion the medium heat, directive of each process, for every fermentable degree, without any regard had to any incident whatever,

Fermentable degrees.	Mean heats to govern the processes.
40	119
45	126

50	133
55	140
60	147
65	154
70	161
75	168
80	175

Now the principal heats affecting common small beer, with regard to its duration, are the degree of heat under which the beer is at first fermented, that of the air when brewed, and when conveyed from place to place, and that of the cellar where it is deposited; let us, in regard to these heats, take the mean of the circumstances this drink is liable to, at the time when the air is at the first fermentable degree, and at the time when the season is hottest (taking for this the medium heat of the whole 24 hours.) Having these two extremes, and making a fit allowance for the hops employed, we shall be able, from the above table, to fix the medium heat that should govern the several processes for making common small beer in every season of the year.

I observed, in page 183, that when the heat of the air is 40 degrees, brewers set the worts of common small beer to be fermented, at a heat of 60; add to this 10 degrees more heat, excited by the fermentable action, makes	70°
The heat of the air we fixed for the first extreme, was the first fermentable heat,	40
In page 156, we said cellars in winter were generally ten degrees hotter than the air, but we observed, those employed for this use, were the worst of the kind, subjected to exterior impressions, or perhaps other defects, for which reason we here set this heat only at	46
Divided by the number of circumstances	3) 156

                                                                                   52°

is the mean of the principal incidents affecting small beer in this season, and, by the foregoing table, this degree indicates a medium to govern the whole process 136, to which must be added, for preservative effect bestowed by the hops used, 1 degree more, which makes it at this heat in the air 137 degrees.

When the mean heat of the whole 24 hours is 60 degrees, (see page 150) if, as in page 183, by the advantage of the evening and night to cool the wort, an abatement of 5 degrees is obtained, the whole of the heat is 55 degrees, add to this only 8 degrees more, because at this time the beer is divided, and put in casks long before the first fermentable act is compleated, and their real heat will be

The medium heat of the air in the hottest season (page 150)	60
In page 156 we say, the heat of the cellars in summer time is generally 5 degrees colder than the exterior air, but these being the worst of the kind, may certainly be thought somewhat more exposed, though not so much affected in summer as in winter, when there are fewer culinary fires, for this reason we fix their heat at	56
Divided by the number of observations	3) 179
	59°

is the mean of these incidents affecting the small beer at this season, and by the foregoing table it indicates a medium heat to govern the whole process 146 degrees, to which, if two degrees more be added, for the effect of the hops, (as experience teaches us six pounds of hops in summer scarcely are so powerful as three pounds in winter) it will give us for the mean of the heats drying the malt, those impressed in the extracts, together with the allowance made for the hops 148 degrees.

Spontaneous pellucidity is always expected in this drink, although the time allotted to gain this in general is much too short; to forward this intent as far as possible, without hazarding the soundness of the drink, in the computations to determine the heats of the first and last extracts, the whole number of constituent parts of malt or 10 degrees are employed.

Having premised these rules, the heats for the first and last extracts are to be found by like operations before made use of, an example of which we shall state; and knowing the mean heats required for two distinct distant processes, in proportion to these I shall form a table, for brewing this drink in every season of the year.

When the air is at 40, the degree of dryness fixed for malts to be used for common small beer is 124, the quantity of hops three pounds per quarter, the medium of their dryness and the heat of the extracts, together with the value of the hops added thereto, is 137 degrees.

124° Malt's dryness.	
137 Medium intended.	
137	
1	Value of hops.
— —	
136	Mean of Malt's dryness, and heat of extracts.
— —	
For the first extract.	
136	As before.
5	Half the number of the whole constituent degrees, to be deducted. (See p. 168.)
— —	
131	
— —	
124	Malt's dryness.
138	Rule to discover the first heat.

— —	
262	
— —	
131	As above.
	For the last extract.
136	As before.
5	Half the number of the whole constituent degrees, to be added. (See p. 168.)
— —	
141	
— —	
124	Malt's dryness.
158	Rule to discover the last heat.
— —	
282	
— —	
141	As above.
— —	
	The proof.
138	Heat of the first extract.
158	Heat of the last extract.
— —	
296	

—	
—	
148	Mean heat of extracts.
124	Malt's dryness.
—	
—	
272	
—	
—	
136	Mean of Malt's dryness and heat of extracts.
1	Value of hops.
—	
—	
137	Medium intended, as above.
—	
—	

The elements for forming common small beer, when the heat of the air is at 40 degrees, independent of the proper division of this heat, adequate to each Mash.

Malt's dryness.	Value of hops.	Whole medium.	First heat.	Last heat.
124	1	137	138	158
			2	2

The medium of the heat lost in the mash ton, amounting to two degrees, is added to the heat of the first and last mash, in the following table.

A TABLE *of the elements for forming common small beer, at every degree of heat in the air, with the allowance of two degrees of heat, in the first and last extractions.*

Heat of air.	Malt's dryness.	Value of hops.	Medium heat of the processes.	First heat.	Last heat.
35	122	1	135	138	158

40	124	1	137	140	160
45	125	1	140	145	165
50	127	1	143	149	169
55	129	1½	146	152	172
60	130	2	148	154	174

From due observation of this table, it appears, how necessary it is for brewers to be acquainted, not only with the daily temperature of the air, but also with the medium heat of such spaces of time, wherein a drink like this is expected to preserve itself. This I have estimated for every 14 days; (page 150) but as the event may not always exactly correspond with our expectations, an absolute perfection in this drink, as to its transparency and soundness, is not to be expected. It greatly depends on the care and attention given to it, and on the temperature and quiescent state of the cellars it is placed in. The first of these circumstances is often neglected, and the other hardly ever obtained, as the places, where common small beer is kept, are generally the worst of the kind.

In keeping beers, every circumstance is assistant to form them so as to obtain elegance in taste, strength, and pellucidity, either spontaneously or by precipitation, but in common small beer; from the shortness of its duration; and from the many complicated incidents that occur; only the medium of the effect of these can be attended to; which governing medium, in general, differs so much from those which form more exact fermentable proportions, that in these extracts, there cannot be expected that near resemblance to natural wines, which, under more favorable management, it is capable of.

The fourth mode of extraction is that, which, by conveying a heat, equal to what is practised for keeping pale strong, and keeping pale small beers, to the liquors commonly known by the names of *pale ale*, *amber*, or *twopenny*, the softest and richest taste malt can possibly yield, and which makes them resemble wines formed from grapes ripened by the hottest sun, though by artfully exciting periodical fermentations, they are, in a very short time, made to become transparent.

As wines have, in general, been named from the town or city, in the neighbourhood of which the grapes, from which they are made, are found growing, this has, though with less reason, been the case, with our numerous class of soft beers and ales. These topical denominations can indeed constitute no real, at least no considerable difference, since the birth-place of any drink is the least of all distinctions, where the method of practice, the materials employed, and the heat of the climate, are nearly the same.

Ales are not required to keep a long time; so the hops bestowed on them, though they should always be of the finest color, and best quality, are proportionably fewer in the winter than in the summer. The reason is, that the consumption made of this liquor in cold weather, is generally for purl[17], whereas, in summer, as it is longer on draught, it requires a more preservative quality.

The properties of this liquor are, that it should be pale; its strength and taste principally depend on the malt, and its transparency should be the effect of fermentation, accelerated by every means, which will not be hurtful to it. Malt capable of yielding the strongest extracts, is such whose dryness does not exceed 120 degrees; and 138 we have seen to be the highest mean of the extracts, and of the dryness of the malt to admit of pellucidity, without precipitation; the hops used, being only so many as are necessary to resist the heat of the seasons the ale is brewed in, may in general be estimated in value, one degree; from these premises, the elements for brewing this drink, will be found by the same rules as before, where 10 degrees are supposed to be equal to the whole of the constituent parts, and the whole of these are employed to accelerate its coming to perfection.

120 Degrees of malt's dryness.

―――

138 Degrees, whole medium intended.

1 Value of hops.

―――

137 Mean of malt's dryness, and heat of extracts.

―――

For the first extract.

137 As before.

5 Half the number of the whole constituent degrees to be deducted.

―――

132 Mean of malt's dryness, and of the heat of first extract.

―――

120 Malt's dryness.

144 Rule to discover the first heat.

264

---

132 As above.

For the last extract.

137 As before.

    5 Half the number of the whole constituent degrees to be added.

---

142 Mean of malt's dryness and of the heat of last extract.

---

120 Malt's dryness.

164 Rule to discover the last heat of last extract.

284

---

142 As above.

The elements for brewing pale ale or amber, with the allowance of 2 degrees for the heats lost in the extracts.

Malt's dryness.	Value of hops.	Medium of the whole.	Heat of first mash.	Heat of last mash.
120	1	138	146	166

The time this liquor is intended to be kept, should entirely be governed by the quantity of hops used therein; for this ale being required to become spontaneously fine, the medium of the whole, or 138 degrees, cannot be exceeded. In and about London, and in some counties in England, these ales, by periodical fermentations, are made to become fine, sooner than naturally they would do, and often, in a shorter time than one week. The means of doing this, by beating the yeast into the drink, as it is termed, has by some been greatly blamed, and thought to be an ill practice. An opinion that the yeast dissolved in the drink, and thereby made it unwholesome, prevailed; and some brewers, erroneously led by this, and yet willing that their commodity should appear of equal strength with such as had undergone repeated fermentations, have been induced to add ingredients to their worts, if not of the most destructive nature, at least very unwholesome. The plain truth is, that, by returning the elastic air in the fermenting ale, the effects of long keeping are greatly imitated, though with less advantage as to flavor and to strength; but as this case relates to fermentation, we shall have hereafter an opportunity of explaining it more at large.

It is under this class, that the famous *Burton ale* may be ranked, and, if I do not mistake, it will be found, that its qualities and intrinsic value will be the same, when judiciously brewed in London, or elsewhere, from whence it may be exported at much cheaper rates to Russia and other parts, than when it is increased in price by a long and chargeable land-carriage.

When drinks are made so strong as these generally are, only two mashes can take place, by which the whole virtue of the malt not being expended, small beer is made after these ales. The purest and most essential parts of the grain being extracted, it is not to be expected, from an impoverished grist, that beers can be made to possess all their necessary constituent parts, or to keep so long, as where fresh malt is used; but the sort of small beer, which answers best to the brewer, and is most salubrious for the consumer, must be, by the addition of fresh hops, to form the remaining strength into keeping small beer, the greater quantity of hops necessary to be allowed, beside those boiled in the ale, is 2¼ pounds for every barrel intended to be made. As much more water must be employed, for this small beer, besides its length, as will steam away in two hours boiling, and 1/8 of a barrel per quarter of malt, for waste. The heat regulating the extract of small, will be found by the following rule.

   138 Medium heat intended for keeping small beer.

    2 Value of hops.

   —
   —

   136 Mean of malt's dryness and heat of extract.

   —
   —

   120 Malt's dryness.

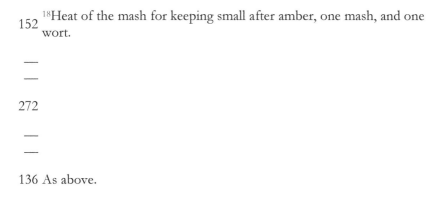

All the hops after these two brewings, as those added for the keeping small beer have been boiled but in one wort, are in value, for the next guile of beer, equal to 1/10 of fresh hops.

We should now put an end to this section, but, as other drinks are brewed besides those here particularly treated of, we shall just mention them, to shew how their different processes are reducible to the rules just laid down.

Brown ale is a liquor, whose length is generally two barrels from one quarter of malt, and which is not intended for preservation. It is heavy, thick, foggy, and therefore justly grown in disuse. The hops used in this, differ in proportion to the heats of the season it is brewed in, but are generally nearly half the quantity of what is employed, at the same times, for common small beer. The system it ought to be brewed upon is not different from that of this last liquor; the medium of the malt's dryness, and heat of the extracts, are the same for each degree of heat in the air, and it requires the same management when under fermentation. But though common pale small beer and brown ale are so much alike in their theory, yet, from the difference of the dryness of the malt, which, for brown ale, is constantly so high as 130 degrees, the practice will appear greatly different. Small beer is made after this ale, by the same rules as that made after pale ale or amber; the malt must, in that case, be valued according to its original dryness, and the medium governing the process be the same as for small beer, and as if no extract had been taken from the grain. No small beer brewed after ales can ever be equal in goodness to such as are brewed from entire grists; but that which is made after brown ale, from the grain being so highly dried, and nearly exhausted, is neither nourishing or fit to quench thirst.

Brown stout is brewed with brown malt, as amber is with pale; the system for brewing these liquors is the same, allowing for the difference in the dryness of the malt. The overstrength of this drink has been the reason of its being discontinued, especially since porter or brown beer has been brought to a greater perfection.—That which is brewed with an intent of being long kept, should be hopped in proportion to the time proposed, or the climate it is to be conveyed to.

Old hock requires the same proportion of hops as are used in keeping pale strong, or keeping pale small beer; but more or less, according to the time it is intended to be kept before it becomes fit for use. The length is about two barrels, from a quarter of the palest and best malt. As spontaneous pellucidity is required, its whole medium must not exceed 138 degrees, for the drying and extracting heat. The management of it, when fermenting, is under the same rules with keeping small beer, or those which are allowed a due time to become of themselves pellucid.

Dorchester beers, both strong and small, range under the same head. They are brewed from barleys well germinated, but not dried to the denomination of malt. The rule of the whole 138 degrees for the governing medium, must, even with this grain, be observed to form these drinks; but, from the slackness of the malt, and the quantities of salt and wheaten flour mixed with the liquor, when under fermentation, proceed its peculiar taste, its mantling, and its frothy property.

SECTION IV.

OF THE NATURE AND PROPERTIES OF HOPS.

THE constituent parts of malt, like those of all vegetable sweets, are so inclined to fermentation, that, when once put in motion, it is difficult to retard their progress, retain their preservative qualities, and prevent their becoming acid. Among the many means put in practice, to check this forwardness of the malt, none promised so much success as blending with the extracts, the juices of such vegetables as, of themselves, are not easily brought to fermentation. Hops were selected for this purpose, and experience has confirmed their wholesomeness and efficacy.

Hops are an aromatic, grateful bitter, endued with an austere and astringent quality, and guarded by a strong resinous oil. The aromatic parts are volatile, and disengage themselves from the plant with a small heat. To preserve them, in the processes of brewing, the hops should be put into the copper as soon as possible, and be thoroughly wetted with the first extract, while the heat of the wort is at the least, and the fire under the copper has little or no effect thereon. Whoever will be at the trouble to see this performed, by the means of rakes, or otherwise, will be made sensible, that flavor is retained, which, when the wort comes to boil, is otherwise constantly dissipated in the air.

The bitter is of a middle nature, or semivolatile: it requires more fire to extract it, than the aromatic part, but not so much as the austere or astringent. Hence it is plain, that the principal virtues of this plant are best obtained by decoction, the austere parts not exhibiting themselves, but when urged by so violent and long continued boiling, as is seldom, or never practised in the brewery. It would be greatly satisfactory to fix, from experiments, the degrees of heat, that first disperse the aromatic, next the bitter, and lastly the austere parts; as it is likely, by this means, a more easy and certain method of judging of the true value and condition of hops, than any yet known, might be discovered.

This vegetable is so far from being, by itself, capable of a regular and perfect fermentation, that, on the contrary, its resinous parts retard the aptness which malt has to this act. Hops, from hence, keep barley-wines sound a longer space of time, and, by repeated and slow frettings, give an opportunity to the particles of the liquor to be more separated and comminuted. Fermented liquors acquire, by this means, a greater pungency, even though it was admitted they received no additional strength from this mixture, the direct contrary of which might easily be made to appear. Hops, then, are not only the occasion of an improvement of taste, but an increase of strength.

Dr. Grew seems to think the bitter of the hops may be increased by a greater degree of dryness; but, perhaps, this is only one of the means of their retaining longer this quality, which undoubtedly decreases through age, in a proportion, as near as can be guessed, of from 10 to 15 per cent. yearly.

The varieties of the soils in which hops are planted, may have some share in the inequality we perceive in them. They seem to be much benefited by the sea air. Whoever will try similar processes with the[19] Worcestershire and Kentish hops, will soon perceive the difference, and the general opinion strengthens this assertion, as the county of Kent alone produces nearly half the quantity of hops used in this kingdom.

The sooner and the tighter hops are strained, after having been bagged, the better will they preserve themselves. The opinion that they increase in weight, if not strained until after Christmas, may be true, but will not recommend the practice; the hops imbibe the moisture of the winter air, which, when the weather grows drier, is lost again, together with some of the more spiritous parts. Nor is this the greatest damage occasioned by this delay, as hops, by being kept slack bagged in a damp season, too often become mouldy.

Hops may be divided into ordinary and strong, and into old and new. The denomination of old is first given to them, one year after they have been bagged. New ordinary hops, when of equal dryness, are supposed to be nearly alike in quality, with old strong ones.

The different teints, with which hops are affected from the fire of the kiln, afford in brewing the best rule for adapting their color to that of the malt; in general the finest hops are the least, but the most carefully, dried.

To extract the resinous parts of the hops, it is necessary they should be boiled. The method of disposing them is generally to put the whole quantity, in the first wort, which, being always made with waters less hot than the succeeding extracts, possesses the greatest share of acids, and is in want of the largest proportion of resins and bitters to defend it. The virtue of the hops is not entirely lost by once boiling, there remains still enough to bitter and preserve the second wort. But where the first wort is short of itself, and a large quantity of hops are required for the whole, it is needless and wasteful to put more in at once than it can absorb, the overplus of which appears by a thin bitter pellicle floating on the wort when laid to cool in the backs. No particular rules can be given to avoid this inconveniency, as the nature and quantity of the worts on one side, and the strength of the hops on the other, must occasion a difference in the management, easily determinable by experience.

When waters, not sufficiently hot, have been used, the wort, for want of the proper quantity of oils, readily admits of the external impressions of the air, and is easily excited to a strong and tumultuous fermentation, which disperses the bitter particles, and diminishes the effects of the hops. The virtue of this plant is therefore retained in the drinks, in proportion to the heat of the extracts, and the slowness of the fermentation.

But beers being a composition of malt, hops, and water, united by heat, and the properties of this combination being judged of by the medium of the whole number of degrees of fire made use of in the process, as we brought the virtues of malt to this denomination, it is also essential to reduce those of hops. After many tedious calculations and experiments, made with this view, and unnecessary here to mention, we were obliged to have recourse to a more simple and probable hypothesis, and confirm the truth thereof by repeated experiments, the relation of which, as it becomes here necessary, will shew the necessity we were under to take a general view of the whole process before we attempted to ascertain this point.

In the table shewing the mean heat of the air applicable to practice, the greatest cold is 35 degrees, and in this season we observed, (page 156) the repositories of beers were more warm than this by 10 degrees, which makes the greatest cold of cellars to be 45 degrees; in the same table the highest heat is 60, when cellars are 5 degrees colder than the external airs, the utmost difference then in the temperature of cellars is 10 degrees, and this takes place in 6 months, so that the whole variety of heat beers deposited for keeping undergo in one twelvemonth is 20 degrees.

There is no specie of beer, in brewing of which it is requisite the artist should be more attentive to alter his process in proportion to the change of heat in the air, than common small beer, which, though brewed in every season, is constantly expected to be in an uniform order for use. In the preceding section, in the table directing this variety, we find a difference of five degrees of heat in the air, requires an alteration in medium heat of the whole process of 3 degrees, and as it is from the mean heat of the dryness of the malt, of the heat of the extracts, and of the value of hops in degrees, that we are to discover the quantity of fire to be given to the extracts, this can be done only by deducting from such medium so much as it is affected by the properties of the hops. Just before we have seen, that the whole of the variety of heat beers deposited in cellars to keep twelve months undergo, amounts to 20 degrees, these, in a proportion of 5 to 3, would be 12, without being scrupulously exact. Hops, with regard to their proportion in the whole process, must be admitted to be one third part thereof, and, in this case, of the proportion, 12, now found, only 4 degrees would be what they contribute towards preserving the drink 12 months: the quantity of hops necessary to maintain beers in a sound state this space of time, we have found to be twelve pounds; this quantity then is equal to 4 degrees of the medium heat of the whole process. On these grounds we repeatedly tried the experiment in a variety of brewings made for different purposes, and never found any inconveniencies from the estimating hops in such like proportion.

Hops should be used in proportion to the time the liquors are intended to be kept, and to the heat of the air in which they are fermented. The quantity requisite to preserve beers twelve months, experience has shewn to be[20] twelve pounds, of a good quality, joined to one quarter of malt, and when the heat of the air is at 40 degrees, three pounds to every quarter has been found sufficient to preserve drinks from four to six weeks, as six pounds are to keep them the same term when the thermometer is so high as 60 degrees. From these facts, founded on informations obtained from long practice, we shall hereafter ascertain the proper quantities to be used in all cases.

Having premised these observations, sufficiently accurate for the government of this art, the construction as well as utility of the following tables will be obvious.

A TABLE *of the value of the hops, expressed in degrees, to be added to the medium of the dryness of the malt, and of the heat of the extracts.*

Hops.	New or strong.	Pale, low dried, or old.
15 lb. equal	5	3¾
12	4	3
8	2	2
4	1	1

A TABLE *of the quantity of hops requisite for every quarter of Malt brewed for porter, supposed to be fit for use from eight to twelve months.*

	lb.
Old ordinary hops started over old beer,	14 per Qr.
Ditto, neat guiles,	12½
Strong good old hops, when started over old beer,	12½
Ditto, neat guiles,	12
New strong hops, when started over old beer,	12
Ditto, neat guiles,	11½

New ordinary hops started over old beer, 12½

Ditto, neat guiles, 12

N.B. The quantity of old beer to be blended with new is here supposed never to exceed one eighth part of the whole.

A TABLE *of the quantity of hops requisite for common small beer, for each quarter of malt, in every season.*

Heat in the air.	New hops. lb. oz.		Old hops. lb. oz.	
35°	2	8	2	8
40	3	0	3	0
45	3	8	3	8
50	4	4	4	8
55	5	0	5	8
60	6	0	6	8
65	6	12		
70	7	8		
75	8	4		
80	9	0		

The medium heat of the hottest days in England, in the shade, seldom, at any time, exceeds 60 degrees, but I continued the table proportionably, as what is here set down is from repeated experiments, and from thence it appears, at the lowest fermentable degree of heat, three pounds of hops are required for each quarter of malt; at the highest, nine pounds of hops should be allowed for the same quantity; this, in some measure, determines the effect of a greater activity in fermentation.

A TABLE *of the quantity of hops necessary to each quarter of malt, in brewing amber or two-penny.*

Heat in the air.	New hops. lb. oz.		Old hops. lb. oz.	
35°	2	8	2	8
40	3	0	3	0
45	3	8	3	8
50	4	0	4	4
55	4	8	4	12
60	5	0	5	4

Amber is a liquor which, by repeated periodical fermentations, is so attenuated, as to be soon fit for use, and, by its strength, is supposed to resist the impressions of the air longer than common small beer, especially in winter; for this reason, it wants fewer hops than that drink does, and in the summer both require equal quantities, on account of the fermentation of amber being carried to a greater degree.

The hops once boiled in amber, but used afterwards for small beer, may be estimated equal to one fourth of their original quality.

When twelve shilling small beer is made after amber, the quality of the hops used should at least be equal in value to the quantity of ten pounds fresh hops to every five barrels of beer, when brewed from entire grists of malt for this purpose.

A TABLE *of the quantity of hops necessary for each quarter of malt, in brewing Burton ale.*

This liquor requires fewer hops than such ales as are more diluted by water: as it is always brewed in the winter, the quantities here set down are for the number of months it is supposed to be kept, before it is drank or bottled.

Months.	lb.	oz.
1	1	0
2	1	8
3	2	0
4	2	8

5	3	0
6	3	8
7	4	0
8	4	8
9	5	5
10	5	8
11	6	0
12	6	8

Though common amber, keeping amber, and Burton ales require the same degree of heat to govern the whole of their processes, yet some small difference will be found in the heats of their extracts, on account of the different quantity of hops used.

Besides the use of hops for keeping the musts of malt, they may also, with great propriety, be employed both to strengthen and preserve sound the extracts. One or two pounds, in a net suspended in the water the mash is to be formed with, are sufficient for this purpose.

Though the purchasing the materials, used in manufacture, does not immediately relate to its practical part, yet as, in this case, it is of great importance to the brewer to know what stock it is prudent for him to keep, of an ingredient equally necessary and variable in its value, I hope the attempt of a calculation on this subject, will easily be pardoned.

The amount of the duty upon hops, for sixteen years, from 1748 to 1765, was £.1,171,227, which sum, estimating the duty at 21*s*. per bag, gives 1,115,454 bags, used in that time. At the beginning and expiration of this interval, hops sold at such high prices, as no considerable stock can be supposed to have remained in hand, viz. from £.8 to £.10 per hundred. If, therefore, to the aforesaid quantity of 1,115,454 bags, which may be supposed to have served for the whole consumption during this period, we add what may have escaped paying duty[21], the annual consumption of hops may be estimated at 70,000 bags, including what is exported to Ireland or elsewhere. From these premises, the following table was constructed, which, though not capable of absolute certainty, may be of some service to the brewers, in informing them of the quantities, that probably remain in hand at any time, and the stock which prudence will suggest to them to lay in.

A TABLE, *shewing the medium price Hops should bear, in proportion to the growth, and determining the quantity to be purchased, in proportion to the stock in hand.*

Prices of hops at a medium[22] per cwt.	Stock of new and old hops in the whole kingdom, after the harvest.	Quantity of hops equal to as many weeks consumption.
30 Shill.	130000 bags,	70
35	125000	65
40	120000	61
45	115000	57
50	110000	53
55	105000	47
60	100000	44
70	95000	40
80	90000	36
90	85000	32
100	80000	28
110	75000	24
120	75000	20
130	70000	16
140	67000	12
150	65000	8
160	62000	4
170	60000	
180	57000	
190	55000	
200	52000	

This chapter should not be dismissed without reminding the brewery, of the gross imposition they submit to in purchasing hops. The tare which justice requires to be allowed in the sale of all packed merchandize, by the hop-factors is refused, who exact payment for the bagging, at the same price as for the commodity itself. If the consumption of hops, in England, is yearly 172,268 cwt. and these be packed one half in bags and the other half in pockets, taking the mean price of hops to be 3l. 14s. per cwt. in this case the consumers are defrauded at least of 39,834l. per annum; that, on a just regulation of this matter, the commodity itself would rise in price, there is not the least foundation for. The present practice of monopolizing hops, by much too frequent, is a farther reason to induce the brewery to exert the influence they ought to have with superior power, to obtain a right so justly due to them.

SECTION V.

OF THE LENGTHS NECESSARY TO FORM MALT-LIQUORS OF THE SEVERAL DENOMINATIONS.

BY length, in the brewery, is understood the quantity of drink made from one quarter of malt. Beers and ales differ in this respect; and the particular strength allowed to every sort of drink, varies also somewhat, according to the prices of the materials. This increase or abatement is, however, never such as to make the profits certain or uniform; for the value of the grain being sometimes double of what it is at other times, a proportionable diminution in strength, can by no means take place.

It might be expected to find here tables determining the differences in strength and quality of each drink, in proportion to their prices, and the expences of the brewer. But this, for many reasons, would be inconvenient, and in some respects impracticable. He, who chuses to be at this trouble, ought not only to take into the account, the prices of malt and hops, but the hazards in the manufacturing them, those of leakage, of bad cellars, and of careless management, the frequent returns, attended with many losses, the wearing out of utensils, and especially of casks, which last article, engrosses at least one fifth of the brewer's capital, the charges of servants, horses, and carriages, for the delivery of the drinks, the duties paid immediately to the government, without any security for the reimbursement, the large stock and credit necessary to carry on this trade, and many other incidents, hardly to be estimated with a sufficient accuracy, and never alike to every brewer. In general it appears, when malt and hops are sold at mean prices, the value of what is employed of these, is equal to the charge attending the manufacture, or of about half the value of the drinks. Hence this conclusion, sensibly felt by every honest trader, that, from change of circumstances, the reputation of the profits has outlived the reality of them, and that a trade, perhaps the most useful to the landed interest, to the government, and to the public, of any, seems distinguished from all, by greater hazards, and less encouragement.

But, in a treatise like this, where only the rules upon which true brewing is founded, are laid down, I would avoid any thing that might, though undesignedly, give handle to invidious reflections, and ill-timed controversies. I therefore content myself with setting down the latitudes of the lengths which should be made for drinks of every denomination.

Lengths of beers, according to the excise gauges, observed within the bills of mortality, or the Winchester measure.

Lengths of common small beer.

4¼ Barrels to 5¼,

Lengths of keeping small beer.

4¾ Barrels to 5½,

Lengths of amber, or pale ale.

1½ Barrel to 1¾, ⎫ from one quarter of malt.

Lengths of brown strong, or porter.

2¼ Barrels to 2¾,

Lengths of Burton ale.

1 Barrel to 1¼,

SECTION VI.

METHOD OF CALCULATING THE HEIGHT IN THE COPPER AT WHICH WORTS ARE TO GO OUT.

THE expected quantities, or lengths of beer and ale, can only be found by determining at what height in the copper the worts must be when turned out.

Brewers have several methods of expressing to what part they would have the worts reduced by boiling. *Brass*, is the technical appellation for the upper rim of the copper; it is a fixed point, from which the estimation generally takes place, either by inches, or by the nails, which rivet the parts of the copper together. These last are not very equal, either in the breadth of their heads, or their distances from each other. Inches then, though not specified on the copper, but determined by the application of a gauge, on which they are marked, claim the preference. The necessity of coppers being gauged, and the contents of what they contain on every inch, both above and below brass, must appear in a stronger light, the nearer we bring the art to exactness. The following tables will shew the most useful manner in which I conceive this gauging should be specified.

Gauges of Coppers.

Great Copper, set up Nov. 30, 1750.   Little Copper, set up Aug. 3, 1753.

		[23]B. F. G.					B. F. G.		
	17	15 3 4	Full			15	11 2 7		
	16	15 2 1				14	11 1 5		
Inches	15	15 0 5	Current	Inches	13	11 0 3	Current
above	14	14 2 8	of	above	12	10 3 1	of
Brass.	13	14 1 4	Great	brass.	11	10 1 7	Little
	12	13 3 7	Copper			10	10 0 6	Copper
	11	13 2 3	allowed.			9	9 3 4	allowed.
	10	13 0 6				8	9 2 2	

	Inches				Inches	
	9	12 3 2			7	9 0 8
	8	12 1 5			6	8 3 6
	7	12 0 1			5	8 2 5
	6	11 2 4			4	8 1 3
	5	11 0 8			3	8 0 1
	4	10 3 3			2	7 2 7
	3	10 1 7			1	7 1 5
	2	10 0 2			Brass	7 0 5
	1	9 2 6			1	6 3 5
	Brass	9 1 1			2	6 2 5
	1	8 3 8			3	6 1 5
	2	8 2 6		Inches	4	6 0 5
Inches	3	8 1 4		below	5	5 1 5
below	4	8 0 2		brass.	6	5 2 5
brass.	5	7 2 8			7	5 1 5
	6	7 1 6			8	5 0 5
	7	7 0 4			9	4 3 4
	8	6 3 3			10	4 2 5
	9	6 2 2			11	4 1 6

By the foregoing table, it is seen that my great copper holds nearly nine barrels of water to brass, and as the difference of the volume between boiling worts, of most denominations, and cold water, is nearly as 7 to 9, the quantity it will yield of boiling worts will be but seven barrels. The diameter of this copper, just above brass, is sixty-eight inches, at a medium, and at that mean it holds twelve gallons seven pints of cold water, or nearly eleven gallons of boiling worts, upon an inch.

Hops macerated, by being twice boiled, take up for every six pound weight a volume, in the copper, equal to four gallons and a half of water, or a *pin*.

In a copper, the gauges of which have just been set down, it is required to know what number of inches a length of twenty-four barrels must go out at, with fifteen pounds of hops, the guile of beer to be brewed at two worts.

    24 Barrels, length of beer.

    14 Barrels, for two full brass,.
    ⎯⎯⎯
    10

     Numbers of gallons to a barrel
    34 accounted by the excise, out of the bills
     of mortality.
    ⎯⎯⎯

    40 Hops twice put in 15lb. is   30.
               ⎯
    30   —.
    ⎯⎯⎯   6lb. [30.
    340   5
    22   4½.
    ⎯⎯⎯   ⎯
Gallons of boiling   Equal to gallons ⎯
wort upon an inch 11 [362   22
    ⎯⎯⎯   ⎯
                ⎯

    33 Inches above brass, the two worts to go out together.

When three worts are boiled, the amount of three full brasses must be deducted from the length; and as the hops go into the copper three times, they become more macerated, and take up much less room. The proportion is then nearly thirteen or fourteen pounds of hops for each four gallons and a half.

Thus in coppers, which have never been tried or used, we are able, by the gauges alone, to determine our lengths; but, as their circumferences are not always exact, and the worts are of very different strengths, we should never neglect such trials as may bring us nearer to accuracy and truth.

SECTION VII.

OF BOILING.

It has been a question, whether boiling is necessary to a wort; but as hops are of a resinous quality, the whole of their virtues are not yielded by extraction; decoction or boiling is as needful as the plant itself, and is, together with extraction and fermentation, productive of that uniformity of taste in the compound, which constitutes good beer.

Worts are composed of oils, salts, water, and perhaps some small portion of earth, from both the malt and hops. Oils are capable of receiving a degree of heat much superior to salts, and these again surpass, in this respect, the power of water. Before a wort can be supposed to have received the whole of the fire it can admit of, such a degree of heat must arise, as will be in a proportion to the quantity of the oils, the salts, and the water. When this happens, the wort may be said to be intimately mixed, and to have but one taste. The fire, made fiercer, would not increase the heat, or more exactly blend together the constituent parts; this purpose once obtained, the boiling of the wort is completed.

It follows from thence, that some worts will boil sooner than others, receive their heat in a less time, and be saturated with less fire; but, as it is impossible, and, indeed, unnecessary, to estimate exactly the quantities of oils, salts, and water contained in each different wort, it is out of our power previously to fix, for any one, the degree of heat it is capable of. This renders the thermometer in this case useless, and obliges us to depend entirely on experiment, and to observe the signs which accompany the act of ebullition.

Fire, as before has been mentioned, when acting upon bodies, endeavours to make its way through them in right lines. A wort set to boil, makes a resistance to the effort of fire, in proportion to the different parts it is composed of. The watery particles are, it is imagined, the first, which are saturated with fire, and becoming lighter in this manner, endeavour to rise above the whole. The salts are next, and last of all the oils. From this struggle proceeds the noise heard when the wort first boils, which proves how violently it is agitated, before the different principles are blended one with another. While this vehement ebullition lasts, we may be sure that the wort is not intimately mixed, but when the fire has penetrated and united the different parts, the noise abates, the wort boils smoother, the steam, instead of clouding promiscuously as it did at first round the top of the copper, rises more upright, in consequence of the fire passing freely in direct lines through the drink, and when the fierceness of it drives any part of the drink from the body of the wort, the part so separated ascends perpendicularly. Such are the signs by which we may be satisfied the first wort, or the strongest part of the extracts, has been so affected by the fire, as to become nearly of one taste. If, at this time, it is turned out of the copper, it appears pellucid, and forms no considerable sediment.

The proper time for the boiling of a wort hitherto has been determined, without any regard to these circumstances; hence the variety of opinions on this subject; greater, perhaps, than on any other part of the process. While some brewers would confine boiling to so short a space as five minutes, there are others who believe two hours absolutely requisite. The first allege, that the strength of the wort is lost by long boiling; but this argument will not hold good against the experiment of boiling a wort in a still, and examining the collected steam, which appears little else than mere water. Those who continue boiling the first wort a long time, do it in order to be satisfied that the fire has had its due effect, and that the hops have yielded the whole of their virtue. They judge of this by the wort curdling, and depositing flakes like snow. If a quantity of this sediment is collected, it will be found to the taste both sweet and bitter, and if boiled again in water, the decoction, when cold, will ferment, and yield a vinous liquor. These flakes, therefore, contain part of the strength of the wort; they consist of the first and choicest principles of the malt and hops, and, by their subsiding, become of little or no use.

It appears, from these circumstances, that boiling a first wort too short or too long a time, is equally detrimental, that different worts require different times, and these times can only be fixed by observation.

The first wort having received, by the assistance of the fire, a sufficient proportion of bitter from the hops, is separated therefrom. The hops, being deprived of part of their virtues, are, on the other hand, enriched with some of the glutinous particles of the malt. They are afterwards, a second, and sometimes a third time, boiled with the following extractions, and thereby divested not only of what they had thus obtained, but also of the remaining part of their preservative qualities. The thinness and fluidity of these last worts render them extremely proper for this purpose. Their heat is never so intense as that of the first, when boiling; for, as they consist of fewer oils, they are incapable of receiving so great a degree of heat. This deficiency can only be made up by doubling or tripling the space of time the first wort boiled, so that what is wanted in the intenseness of heat, may be supplied from its continuance.

The following table is constructed from observations made according to the foregoing rules.

A TABLE *shewing the time each wort requires to boil for the several sorts of beer, in every season.*

	Brown beer, keeping pale strong and keeping small beer.			Small beer.			amber	Burton	small after amber	keeping small after amber
	hours	hours	hours	hours	hours	hours	hours	hours	hours	hours
35°	1	2	4	1/2	1	2	1/2	1/2	1	2
40	1	2	4	1/2	1	2	1/2	1/2	1	2
Degrees of heat in the air 45	1	2	4	1/2	1	2	1/2	1/2	1	2
50	1	2	4	1/2	1	2	1/2	1/2	1	2
55		2	4		1½	3	3/4	3/4	1	2
60		2	4		1½	3	3/4	1	1½	2

1 wort 2 wort[24] 3 wort 1 wort 2 wort 3 wort.

It may, perhaps, be objected, that, by a long boiling of the last worts, the rough and austere parts of the hops may be extracted, and give a disagreeable taste to the liquor; but it should be observed, this only happens, either in beers to be long kept, or in such as are brewed in very hot weather. In the first case the roughness wears off by age, and grows into strength, and in the last, it is a check to the proneness musts have in such seasons to ferment.

One observation more is necessary under this head; most coppers, especially such as are made in London, and set by proper workmen, waste or steam away, by boiling, about three or four inches of the contained liquor, in each hour. The quantity wasted being found on trial, and knowing how much water the copper holds upon an inch, what is steamed away by boiling in each brewing, may easily be estimated.

SECTION VIII.

Of the Quantity of Water wasted; and of the Application of the preceding Rules to two different processes of Brewing.

WASTE water, in brewing, is that part which, though employed in the process, yet does not remain in the beers or ales when made. Under this head is comprehended the water steamed away in the boiling of the worts; that which is lost by heating for the extracts; that which the utensils imbibe when dry; that which necessarily remains in the pumps and underback; and more than all, the water which is retained in the grist. The fixing to a minute exactness how much is thus expended, is both impossible and unnecessary. Every one of the articles just now mentioned varies in proportion to the grist, to the lengths made, to the construction and order of the utensils, and to the time employed in making the beer. To these different causes of the steam being lessened or increased, might be added every change in the atmosphere. However, as, upon the whole, the quantity of water lost varies from no reason so much, as from the age and dryness of the malt, experience is, in this case, our sole and surest guide. I have, in the following table, placed under every mode of brewing, how much I have found necessary to allow for these several wastes and evaporations.

Brown strong and pale strong beers.

	Barrels	pins.[25]	
For old malts allow	1	5	per quarter.
For new[26] malts	2	0	per quarter.

Keeping small and common small beers.

For either new or old malt allow	2	4	per quarter.

Amber or pale ales.

For either new or old malt allow	1	5	per quarter.

Keeping small or common small after amber.

Allow for waste	0	2	per quarter.

It is now time to begin the account of two brewings, which admit of the greatest variety, both in themselves, and in the season of the year. The same processes will be carried on, in the sequel of this work, until they be completed.[27]

On the tenth of July a brewing for common small beer is to be made with 6 quarters of malt.

By page 150 the medium heat of the air at this time is } 60 degrees.

By page 184 the malt to be used for this purpose should be in dryness at } 130 degrees.

By page 210 the proper quantity of new hops is 6 pounds per quarter. The length, according to the excise gauge without the bills of mortality, may be rated at 5 barrels 1/8 per quarter, or from the whole grist at 30 barrels 3/4. See page 219.

By page 222, the inches required in the copper, to bring out this length, at 2 worts, will be, for coppers as gauged page 221, 56 inches in the 2 worts above brass.

The state of this part of the brewing is, therefore, six quarters of malt dried to 130 degrees, 36 pounds of hops for 30 barrels 3/4 to go out at 56 inches above brass.

30¾ Length

    Boiling by page 228

      { 1 wort 1 hour 1/2 or 5 inches.

5¼   2 wort 3 hours or 9 inches.

15   waste water page 231

———

51 barrels; whole quantity of water to be used.

And by page 191 we find the heat of the first extract to be 154 degrees, and the heat of the last 174 degrees.

The other brewing, of which I purpose to lay down the process in this treatise, is one for brown beer or porter of 11 quarters of malt, to be brewed on the 20th of February.

By page 150 the medium heat of the air at this time is } 40 degrees.

By page 174 the malt to be used for this purpose should be at } 130 degrees.

By page 209 the quantity of hops is 12 pounds per quarter. The length I would fix for this liquor, according to the excise gauge without the bills of mortality, is 2 barrels and 4 pins from a quarter, or from the whole grist 27 barrels 1/2. See page 219.

By page 222, the inches required, in a copper, such as I have specified page 221, to bring out this length at 3 worts, are 31 above brass.

The state of this brewing, so far as we have considered it, is therefore 11 quarters malt dried to 130 degrees, 132 pounds of hops for 27 barrels 1/2 to go out at 31 inches above brass.

27½   barrels the length,

     Boiling by page 228

     1 wort 1 hour or 4 inches.
    {
     2 wort 2 hour or 6 inches.

8¼   3 wort 4 hours or 12 inches.

18   waste water page 231 old

――   malt 1-5/8 per quarter.

54   barrels; whole quantity of water to be used.

And by page 177 we find the heat of the first extract to be 155 degrees, and the heat of the last extract 165.

SECTION IX.

Of the Division of the Water for the respective Worts and Mashes, and of the Heat adequate to each of these.

THAT the whole quantity of water, as well as that of heat required, ought not, in any brewing, at once to be applied to the grist, is obvious, both from reason, and from the example of nature, who, in forming the juice of the grape, divides the process, and increasing successively both the moisture and the heat, gives time to each degree to have its complete effect. A division of the water and heat to form malt liquors is equally necessary, but previous to this division the following general rules may be laid down.

The grist, if possible, is at no time to be left with less water than what will cover the malt, to put all its parts in action. In the first mashes for strong beer, an allowance is to be made for nearly as much water as the grist will imbibe; and, lastly, the whole quantity of water used in brewing should be divided, in a proportion analogous to that of the degrees of heat.

Processes for brewing are carried on either with one copper or with two. Though the first of these methods is almost out of use, it may be necessary to give an example or two of the division of the water used in this case, the doing which will point out the absurdity of this practice.

In brewing with one copper, scarcely more than three mashes can be made; otherwise the time taken up in boiling the worts, and preparing the subsequent waters for extraction, would be so long, as to cause the grist to lose great part of its heat, and, in warm weather, perhaps, to become sour. The whole water required might naturally be divided into three equal parts, was it not for the quantity at first imbibed by the grist; but as, in this way of brewing, the best management is to make the first wort of one mash, and the second wort of the other two, it will be found necessary to allow, for the first extracting water, four parts out of seven of the whole quantity required, and to divide the remainder equally for the other two mashes. Thus, if the whole quantity of water required was fifty-one barrels, the lengths of the extracting waters would be as follow:

1 Liquor	2 Liquor	3 Liquor
29	11	11 Barrels.
1 Wort.	⌒_____v_____	
	2 Wort.	

The water imbibed and retained by the malt is allowed for in this computation, which will be found just to every purpose, for small beer brewed in one copper only.

But in strong beers and ales, with three mashes, whether brewed at one, two, or three worts, the case will be somewhat different, as care should always be taken to reserve for every mash a sufficient quantity of water to apply to the grist. For this reason, no greater proportion ought to be used in the first mash than that of three parts out of seven, as the volume of the malt is in a greater proportion to the quantity of water than in the preceding case. If, therefore, the whole quantity of water used was thirty-five barrels, the length of the liquors would be:

1 Liquor	2 Liquor	3 Liquor
15	10	10 Barrels.

Employing only one copper, must from hence appear, and is allowed to be, bad management; for, in some part or other of the process, however well contrived, the business must stand still, and consequently the extracts be injured, by the air continually affecting them. The best and most usual practice, and that which here will be set in example, is to brew with two coppers. Other rules consequently are necessary to be observed, and I shall be more particular in the explanation of them.

To preserve order, and to convey our ideas in the clearest manner, we shall make use of the four modes of brewing we mentioned, in the fourth section.

The first of these, which implies keeping pale strong and keeping pale small beers to become spontaneously fine, are best brewed with two worts and four mashes, to allow for what is imbibed by the grist, and what is steamed away during the first part of the process, four sevenths of the whole of the water employed, and consequently a like proportion of the number of the degrees which constitute the difference between the first and last heats of the whole brewing, are required for the first wort, and the remainder to the last or second. The proportion as to the water is permanent, but having now only a division of heat in a progressive state, for the temperature to be given to the extracts, to put in practice the principles laid down in pages 64, 65; the first wort, however, composed of several mashes, must be of one uniform heat, though less than that of the second, whose extracts, though more powerful, must, notwithstanding, be of equal heat among themselves.

According to the rules laid down in section 8, the whole quantity of water requisite for a guile of keeping pale strong, or keeping pale small beer, is fifty-one barrels. In page 171, we found, including the heat lost at the time the extract separates from the grist, the first heat to form this process to be 144 degrees, and the last 158 degrees; the quantity of water, and the difference between these two degrees, are required to be divided in such proportions as are best applicable to the purpose we intend.

          Water   51 Barrels, multiplied by

                                4
                              ———

Divided by 7)   204

Gives   29 Barrels for the first Wort, and this deducted from 51,

Leaves   22 Barrels for the second Wort.

The twenty-nine barrels, equally divided between the two first mashes, is fourteen barrels and a half for each; and the twenty-two barrels, equally divided between the two last mashes, is eleven barrels for each.

The last heat for pale keeping beers is	160 degrees.
And the first is	146 degrees.
Their difference is	14
This, as above, multiplied by	4
And divided by	7) 56
Leaves	8 degrees.

the proportion to be allotted to the first wort, and 6 degrees, the remainder, to the last, in a regular progressive state; the elements for this brewing would stand as under.

	Malt's dryness.	Value of hops.	Whole medium.	First mash.	Second mash.	Third mash.	Fourth
Degrees	119	3	138	146	154	157	160
Barrels				14½	14½	11	11

But more exactly, to imitate the fermented liquors formed by nature, our first wort, answering to the germinating part of her process must be of one uniform heat in the extracts, as must likewise our second wort: (See page 165) the mean, then, of the progressive heats of the first wort will be that which must be applied both to the first and second mashes, and the mean of the progressive heats of the second wort, that which must direct the third and fourth mashes; from whence are deduced

Elements for forming keeping pale strong and keeping pale small beers.

	Malt's dryness.	Value of hops.	Whole medium.	First mash.	Second mash.	Third mash.	Fourth mash.
Degrees	119	3	138	150	150	158½	158½
Barrels				14½	14½	11	11

$$\underbrace{\qquad\qquad}_{\text{First wort.}} \quad \underbrace{\qquad\qquad}_{\text{Second wort.}}$$

That this method of applying the heats to the mashes corresponds to the medium heat which is to govern the whole process, the circumstances required in page 165, the following operation will prove.

29 Barrels, the first wort.

Heated to 150
———
1450
29
———
4350

                    ─────
                    22 Barrels, the second wort.

Heated to 158½
                    ─────
                        11
                       176
                       110
                        22
                    ─────
whole   3487
quantity   4350
of water   ─────
Barrels 51)7837(153 The mean heat of the 4 mashes.
            51   2 Deducted for the heat lost at the tap.
          ─── ───
           273   151 Heat of the tap's spending.
           255   119 Malt's dryness.
          ─── ───
           187   270
           153   ───
          ───   135 Mean heat of Malt's dryness and of the extracts.
                      3 Value of hops.
                    ─────
                    138 Mean heat of the whole process.
                    ─────

Admitting of the necessary variations in the medium heats which are to govern processes for different purposes, and of those in the number of degrees forming the constituent parts of the must, in proportion as the drinks are to be formed, either to become spontaneously fine, or made so by precipitation, or intended for a longer or shorter duration. This rule will be found universally true, when beers are brewed with two worts: but when, for the benefit of the drink, or on account of the smallness of the utensils, as is often the case, when the second mode of extraction is put in practice, we are obliged to carry on the process with three worts, these proportions must necessarily be altered, and the following have, in this case, been found most advantageous.

The first and second wort ought to have two thirds of the water; the first wort two thirds of this quantity, the second the remainder of this, and the third wort one third part of the whole.

Porter or brown beer is the sort of drink, in which this division is most commonly observed. Let the whole quantity of water to be used be that of the brewing, of which the elements have been laid down, (page 233) or 54 barrels.

$$
\begin{array}{r}
54 \\
2 \\
\hline
3)\,108 \\
\hline
36 \\
2 \\
\hline
3)\,72 \\
\hline
\end{array}
$$

24 Barrels of water for the first wort.

12 Barrels of water for the second wort.

18 Barrels of water for the third wort.

      —
      54
      —

The last degree for this drink is, with malt dried to 130 degrees,	165 Degrees.
The first, as per page 178	155 Degrees.
Their difference	10 Degrees.

      2
      —
   3) 20
      —
      7
      2
      —
   3) 14
      —
   5 Heat of first wort.

Five degrees to be proportioned in the first wort, and these deducted from 7 degrees, the number allowed for the first and second wort, there remains two degrees for the second wort; and seven degrees deducted from ten, the whole difference, leaves three degrees, to be proportioned in the third and last wort.

A grist of eleven quarters of malt is too large, to admit of the water allowed for the first wort to be equally divided between the first and second mash; therefore, rather than use the whole 24 barrels in one mash, a sufficient quantity only must be applied to the first mash, both to work it, and to get as much of the extract to come down, as will save the bottom of the copper it is to be pumped into. By this management, there will be enough left to form the second extract with, or what by the brewers is termed the piece liquor. The exact quantity of water the first mash should have, might be referred to the following section, but the order we have laid down, will excuse our anticipating thereon.

It has been found, and will hereafter be proved, that a volume of eleven quarters of malt, dried to 130 degrees, is equal to 6,32 barrels of liquid measure, that malt in general requires twice its volume of water to wet it, and this quantity of water is retained after every tap is spent.

6,32 Barrels, volume of the 11 quarters of malt.

3

―――

18,96

6,32

―――

12,64 Barrels of water imbibed by the grist, which, deducted from

24,00 Whole quantity of water allowed for the first wort.

―――

Remains 3) 11,36 Extract, which will be yielded from the first and second mash.

3,78 Length of the first piece, which is sufficient to save the copper.

―――

3,78

12,64 Quantity imbibed as above.

―――

16,42 Quantity of water for the first mash.

7,58 Quantity of water for the first mash.

———

24,00

The elements of this brewing, as we have them (page 178) placed in a progressive state, will be as under, where the quantity of water allowed for the first wort is divided into two mashes, according to the circumstances just now taken notice of, where the second wort is formed by one entire mash, and the water allotted for the third wort is separated equally into two parts, for the two last mashes, and when the ten degrees of heat, the difference between the first and last heats employed, are as near as possible proportioned to the lengths of the worts.

	Malt's dryness.	Value of hops.	Whole medium.	First mash.	Second mash.	Third mash.	Fourth mash.	Fifth mash.
Deg.	130	4	148	155	160	162	164	165
Barrels				16	8	12	9	9

But, for the reasons alleged in page 236, they admit of the following variation.

Elements for brewing brown beer or porter.

	Malt's dryness.	Value of hops.	Whole medium.	First mash.	Second mash.	Third mash.	Fourth mash.	Fifth mash.
Deg.	130	4	148	157½	157½	162	164	165
Barrels				16	8	12	9	9

Groupings: (First mash + Second mash) = 1 wort; (Third mash) = 2 wort; (Fourth mash + Fifth mash) = 3 wort.

And, if proved as before, the same correspondence will be found with the medium governing heat.

The third mode of extraction is intended for a drink which is soon to be ready for use, in which, in the coldest season of the year, transparency is expected, and, in the hottest months, soundness: to procure these intents, we have already shewn (page 191) it was necessary to vary the medium heats governing these several processes, in proportion as the seasons of the year differed as to heat and cold. Our present business is a proper division of the whole quantity of water necessary for brewing, into the respective worts and mashes, and to apply to each, the adequate degree of heat: one single example will suffice for the operation, and the whole variety this drink is subjected to, will be expressed in the table subjoined.

The general practice to brew common small beer, and which is best, is to form it with two worts and four mashes, and, in this case, as was before practised for keeping pale beers, in order to allow for the water at first absorbed by the grist; four sevenths of the whole quantity is required for the first wort, and the remainder for the second wort, dividing these quantities again into equal parts, for their respective mashes. As a speedy spontaneous pellucidity is expected in every season of the year, and as every means for producing this without affecting the soundness of the drink, must be put in practice, the whole number of constituent parts are not only applied, but likewise the progressive heats suffered to take place: for here, through necessity, we are compelled to forsake the rules nature pointed out, (as in pages 64, 65); the reasons why are obvious; this drink receives no benefit by the slow progress nature recommends, and therefore very little by the impressions of time.

In page 232, we found the whole quantity of water to be used for the brewing there specified, fifty-one barrels, and in page 191, we find when the heat of the air is at 60, the first heat is 154, the last 174 degrees.

       Water   51 Barrels, multiplied by

                       4
                      ———

Divided by 7)   204
                   ———

       Gives   29 for the first Wort, and this deducted from 51,

       Leaves   22 for the second Wort.

The twenty-nine barrels, divided into the first and second mashes, will be fourteen barrels and a half for each; and the twenty-two barrels, equally divided between the third and fourth mashes, is eleven barrels each.

The last heat for this brewing of common small beer is (see page 191)	74 degrees.
The first heat,	154 degrees.

Their difference	20
Multiplied by	4
And divided by	7) 80
Leaves (to avoid fractions) nearly	12

to be proportioned in the first wort, and 8 degrees, the remainder of the 20, to the second wort, in a regular progressive state: the elements for this brewing are:

	Malt's dryness.	Value of hops.	Whole medium.	First mash.	Second mash.	Third mash.	Fourth mash.
Deg.	130	2	148	154	166	170	174
Barrels				14½	14½	11	11

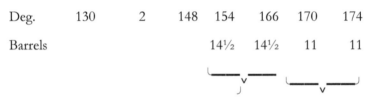

First wort.　　Second wort.

The quantity of water used for brewing small beer is in proportion to the largeness of the grist, and the price of the grain; this admitting of almost an endless variety, it is needless to pursue it: but the dryness of the malt, the value of the hops, the medium governing the processes, and the heat of the extracts being fixed, and constant degrees of heat in proportion to that of the air, I have constructed the following table, which will be found useful to the practitioner in every season of the year.

Heat of air.	Malt's dryness.	Value of hops.	Whole mash.	First mash.	Second mash.	Third mash.	Fourth mash.
35	122	1	135	138	150	154	158

40	124	1	137	140	152	156	160
45	125	1	140	145	157	161	165
50	127	1	143	149	161	165	169
55	129	1½	146	152	164	168	172
60	130	2	148	154	166	170	174

The last business of this section is to divide the quantity of water requisite to brew pale ales or amber, and to apply to such divisions their necessary degrees of heat. This liquor is rather an effort of art, than an exact imitation of nature, as in it the greatest transparency, joined to the greatest strength, is expected in a very short time. To obtain these ends, the whole number of the constituent properties of malt and two mashes only are employed. In the first, in order to favor its pellucidity, the lowest adequate extracting degree must be used; and in the second, to cause the malt to yield the whole of its necessary parts, the highest fitting heat must be applied; the whole of the process is, nevertheless, subjected to the governing medium heat of 138 degrees, the highest which admits of voluntary brightness. But where a drink is formed with two mashes only, and boiled off in one entire wort, to keep the due proportion between the quantity of water used, and the heat required in the extracts, and at the same time to allot the proper quantity for what is imbibed by the grist, the most convenient division found, will be three-fifths of the whole quantity of water to be applied to the first mash, and the remaining two-fifths to the other. I know to this, custom may be objected, that the first mash for amber should be a stiff one, in order the better to retain the heat; but this, in the division here proposed, may equally be obtained by a proper allowance made in the attemperating of the water, without affecting the proportion of the heats required, as otherwise must be the case.

From 8 quarters of malt to make 13 barrels of fine ale.

       13 Length.

       ½ Boiling half hour.

       12½ Waste water.

       ———

       26 Whole water employed, multiplied by

$$\begin{array}{r}3\\ \text{Divided by 5)}\,78\\ \hline\end{array}$$

Gives   16 Barrels for the first mash, and leaves

10 Barrels for the second mash,

the lowest heat being required in the first extract, and the highest in the last, according to page 194; for the 16 barrels it will be 144, and for the 10 barrels it will contain 164 degrees.

But as the heat of the air occasions a difference in the quantities of hops to be used, and as from hence the extracts are somewhat varied: it has been judged convenient to add the following table:

A TABLE *of the elements for forming pale ale or amber, at every degree of heat in the air, with the allowance of two degrees of heat, in the first and last extractions.*

Heat of air	Malt's dryness	Value of hops.	Medium heat of the extracts, and of malt's dryness.	First heat.	Last heat.
35	120	½	138	147	167
40	120	¾	138	146	167
45	120	1	138	146	166
50	120	1½	138	145	165
55	120	1½	138	145	165
60	120	2	138	144	164

In summer time, it is sometimes thought better to brew this drink with malts more dried; for conveniency sake, I here insert two examples.

Heat of air.	Malt's dryness.	Value of hops.	Whole Medium.	Heat of first mash.	Heat of last mash.
60	122	2	138	142	162
60	124	2	138	140	160

For the management of small beer made after amber, see page 197.

Thus having shewn how to ascertain the quantities of the malt, the hops, the water, and the heat to be used, and to proportion them to each other, as the good or bad properties of beers arise from the extracts, and fire is the governing agent, we must now seek the means to administer the right portion of heat, and so to temper the water that is to form the extracts, as not to be disappointed of our intentions. In the calculations made for this purpose, not only the water in the copper, but the value and effect of the grist, as to heat and cold, must be considered.

SECTION X.

An enquiry into the Volume of Malt, in order to reduce the Grist to liquid Measure.

THE gallon, by which malt is measured, though less, is nearly of the same capacity with that, which is used for beer or water. The quarter of malt, contains 64 gallons of this measure, and the barrel, within the bills of mortality, according to the gauges used by the excise, contains 36 gallons, but without the bills, 34; though the first quantity is the measure for sale throughout the kingdom. Hence it would appear, that proportioning the grain to the barrel of water would be no difficult undertaking. This however is so far from being the case, that, after having made use of several calculations to help us to the true proportions, we shall find, they want the corroborating proofs of actual experience, to be entirely depended upon.

The ultimate parts of water are so very small, as to make this, as well as all other liquids, appear to the eye one continued uniform body, without any interstices. This cannot be said of malt laying together either whole or ground; there are numbers of vacancies between the corns, when whole, and between the particles when ground, but for our present purpose the volume occupied by any quantity of malt is properly no more, than the space which would be occupied by every individual corn, either whole or cut asunder, were they as closely joined together as water.

To determine, with precision, the quantity of cold water to be added to that, which is brought to the boiling point, (an act by the brewers called *cooling in*) it is necessary to know, what proportion a quarter of malt bears to the measure of a barrel of water. Several operations will be found requisite to come to this knowledge; viz. to take several gauges of different brewings, more especially in the first part of the process; to be well acquainted with the degree of dryness of the malt used, the heat of the first extract, and the quantity of liquor the mash tun holds upon every inch; to find out what degrees of expansion are produced by the different degrees of heat in the first mash, how much less water the mash tun holds upon an inch when hot, than it does when cold, what quantity of water is lost by evaporation, and in what proportion at the several terms of the process. In order to put this in practice, the gauges of the following brewings were taken.

5 quarters of malt dried to 125 degrees.

B^{28}. F. G.

The quantity of water used for the first mash was set	} 12	2	3

The malt and water gauged together in the mash tun just before the tap was set	} 25, 00 inches.
Allowance for the space under the false bottom boards of the mash tun, as near as could be computed	} 0, 66 inches.
The goods gauged in the mash tun, after the first tap was spent	} 15, 41 inches.
	B. F. G.
First piece gauged in the copper	8 0 2

	B. F. G.
The water employed for the second mash was	} 12 2 3
The grist gauged with this water just before the tap was set	} 30, 62 inches.
And just after the tap was spent	15, 63 inches.
	B. F. G.
The first wort consisting of these two pieces gauged in the copper	} 21 2 0

	B. F. G.
The water used for the third mash was	8 3 6
Just before the tap was set the grist with this gauged in the mash tun	} 24, 60 inches
And just after the tap was spent	15, 20 inches.

	B. F. G.
The water used for the fourth mash was	8 3 6
The mash gauged just before the tap was set	} 24, 60 inches.
And just after the tap was spent	15, 16 inches.

The heat of the first extract was 136 degrees, to which adding two degrees, for what is lost by the tap spending, the true heat of the mash is 138 degrees.

The first extract, before it is blended with hops, may be estimated to be nearly as strong as a first wort of common small beer. This, when under a strong ebullition, raised the thermometer to 216 degrees, and seven barrels of such a wort, when boiling, occupied an equal space with nine barrels of cold water, at the mean temperature of 60 degrees. Now, if the degrees of expansion follow the proportion of those of heat, the following table, constructed upon this supposition, will shew how many barrels of cold water would be necessary to occupy the same space with seven barrels of wort of different heats.

Degrees of heat.	Barrels of cold water.	Barrels of wort.
216	9,00	7
206	8,87	7
196	8,75	7
186	8,62	7
177	8,50	7
167	8,37	7
158	8,25	7
148	8,12	7
138	8,00	7
127	7,87	7
119	7,75	7

The quantity of water evaporated in a brewing, when not in immediate contact with fire, is more considerable than it is generally apprehended to be; after repeated trials, I have found that what was lost in this manner amounted nearly to one fifth.

Now since the heat of the first tap was 138 degrees, and my mash tun holds 20,25 gallons upon an inch, the following proportion may be deduced from the preceding table.

If 8 ——————7—————— 20,25

8,00) 141,7500

17,71 Gallons,

and this is the true quantity contained in one inch, at a heat of 138 degrees.

The quantity of water used for the first mash, was 12B. 2F. 3G. or 428 gallons, of which one fifth is supposed to be steamed away, when the first liquor is gone through the whole process of the extraction: but as the gauges of the malt and water together are taken before the tap is set, in the beginning of the process, the whole evaporation ought not to be deduced, and one sixth seems to be a sufficient allowance on this account. We may therefore suppose 357 gallons to be in the mash tun at the time of gauging, which number being divided by 17,71, will shew how many inches are taken up by the water at that heat.

$$17,71)357,0000(20,15$$
$$3542$$

$$2800$$
$$1771$$

$$10290$$
$$8855$$

$$1435$$

The mash gauged just before the tap was set,	25,00 Inches.
Allowed for the space under false bottoms,	0,66
	25,66
Deduct the inches taken up by the water,	20,15
Remainder for the five quarters of malt,	5,51 Inches.

or 1,10 inch for one quarter. This number being multiplied by 17,71, the quantity of gallons contained upon one inch at this heat, will give 19,48 gallons for the volume of one quarter of this malt. There now remains nothing but to bring a barrel of water of 34 gallons, under like circumstances, as to expansion and evaporation, with these 19,48 gallons, with this difference only, that as the proportion required is, at the time the water and malt first come in contact, and not after the mash has been worked, a less allowance for steaming will be sufficient, and may well be fixed at one seventh.

Gauge within the bills of mortality.	Gauge without the bills of mortality.
If 7,00 8,00 36	If 7,00 8,00 34
36	34
———	———
4800	3200
2400	2400
———	———
7,00)288,00	7,00)272,00
———	———
41,14	38,85
5,87 Lost by steam.	5,55 Lost by steam.
———	———
35,24	33,30

The barrel of water reduced; and as 19,48 gallons, under the same circumstances, were found equal to one quarter of malt, the following division will shew the proportion, between them.

19,48)35,2400(1 81	19,48)33,3000(1,70
1948	1948
———	———
15700	13820

                15584   13636
             ─────────   ─────────
                 1760   184
                 1948
             ─────────

Thus, in malt dried to 125 degrees, the quantity of 1,70 quarters is required to make a volume equal to 34 gallons, or a barrel of water, according to the excise gauging without the bills of mortality; and the quantity of 1,81 quarters is required to make a volume equal to 36 gallons, or a barrel of water, according to the excise gauging within the bills of mortality.

The more the malt has been dried, the larger the interstices are between its parts; the quantity of water it admits will consequently be greater than what is absorbed by such as is less dry. More of this last malt will be necessary to make a volume, equal to that of the barrel of water; and every different degree of dryness must cause a variety in this respect. It will therefore be proper to repeat the operation with a high-dried grist.

Gauges of a brewing of eight quarters of malt dried to 140 degrees.

	B. F. G.
The quantity of water used for the first mash,	} 11 2 4
The malt and water gauged together in the mash tun just before the tap was set,	} 26,25 Inches.
Allowed for the space under the false bottom of the mash tun,	} 0, 66 Inches.
Goods gauged in the mash tun after the first tap was spent,	} 22,36 Inches
	B. F. G.
First piece gauged in the copper,	5 0 0

	B. F. G.
The water for the second mash was	11 2 4

The mash gauged just before the tap was set,	35,70 Inches.
Just after the tap was spent,	22,19 Inches.
	B. F. G.
The wort made of these two pieces gauged in the copper, }	17 0 0

	B. F. G.
The water used for the third mash was	8 3 6
The mash gauged just before the tap was set	31,10 Inches.
And just after the tap was spent,	21,77 Inches.

	B. F. G.
The water used for the fourth mash was	8 3 6
The mash gauged just before the tap was set	30,50 Inches.
And just after the tap was spent	21,60 Inches.

The heat of the first extract was 142 degrees. Now, by the table of expansions (page 256).

		G.	
If 8,05	7,0	20,25	of cold water, upon
		700	an inch in mash tun.

		8,05)1417500(	17,60 will be the real
		805	quantity of water.
		---	upon an inch in the
		6125	mash tun, when heated
		5635	to 142 degrees.

		4900	

	4830	
	—	
	700	

	B. F. G.
Quantity of water in the first mash,	11 2 4
	34
	—
	44
	33
	17
	4
	—
	395
Deduction for the evaporation at this period, one sixth,	65,83
	—
	329,17 true quantity

of the water for the first mash, which must be divided by the real quantity of water contained upon an inch in the mash tun.

17,60)329,1700

    1760   (18,70 inches taken up
    ———   in the mash tun, by
    15317   the water used in
    14080   the first mash.
    ———

```
    12370
    12320
    ─────
       50
```

The mash gauged just before the tap was set	26,25 Inches.
Allowed for the space under the false bottoms	0,66
	─────
	26,91
Inches taken up by the water of the first mash	18,70
	─────
Space occupied by these 8 quarters of malt	8) 8,21 Inches of
	───── mash tun.
Space occupied by one quarter	1,02
	17,60
	─────
	6120
	714
	102
	─────
	17,9520 Gallons of

water equal in volume to one quarter of this malt.

Excise gauge without the bills of mortality.

                    If 7,00   8,05 34

                                    34
                                    ———
                                    3220

                                    2415
                                    ———
                                    7,00)273,70
                                    ———

                                    39,10 Expansion of the barrel of

water out of which 1/7th,   5,58 is to be deducted for

evaporation.   ———

Remains,   33,52 for the barrel of water

reduced, which the quarter of malt, or 17,95, is to be compared to.

Excise gauge within the bills of mortality.

If 7,0   8,05   36

                 36
                 ———
                 4830

                 2415
                 ———
                 7,00)289,80 (41,44 Expansion of one barrel of water,

                 2800

                 ——— 592 1/7th to be deducted for evaporation.

- 161 -

$$
\begin{array}{r}
980\text{ ———}\\
700\\
\text{———}\\
2800\\
2800\\
\text{———}
\end{array}
$$
35,52 Barrel of water reduced, which the quarter of malt, or 17,95 is to be compared to.

$$17{,}95) 33{,}5200 (1{,}86$$

Quantity of malt dried to 140 degrees equal to one barrel of water.

$$
\begin{array}{r}
1795\\
\text{———}\\
15570\\
14360\\
\text{———}\\
12100\\
10770\\
\text{———}\\
1330
\end{array}
$$

$$17{,}95) 35{,}3700 (1{,}97$$

Quantity of malt dried to 140 degrees, equal to one barrel of water, according to the excise gauge within the bills of mortality.

$$
\begin{array}{r}
1795\\
\text{———}\\
17420\\
\text{———}\\
12650\\
12565\\
\text{———}
\end{array}
$$

Having found the volume of malt at two distant terms of dryness, we might divide the intermediate degrees in the same manner as we have done before, could the certainty of these calculations be entirely depended upon; but as some allowances have been made without immediate proof, how near soever truth the result thereof may from experiments appear, it may be proper to point out what is wanting to make our suppositions satisfactory.

Some part of the calculation depends on the quantity evaporated; this, in the same space of time, may be more or less, as the fire under the water is brisk or slow, or as the weight of the atmosphere differs. The gauges are taken at the time the malt and water are in contact, and more or less water may be imbibed in proportion, both of the dryness and age of the malt; water as a fluid, malt as a porous solid body, must differ in their expansion, but in what proportion is to me unknown; effervescence may be another cause of want of exactness; the different cut the malt has had in the mill, its being or not being truly prepared, and lastly the difference as to time, of the mashing or standing of the grist, prevent our relying wholly upon the calculation. It is, however, not improbable that some of these incidents correct one another. Since 1,70 quarter of malt dried to 125 degrees are equal to one barrel of water, and 1,86 quarter of malt dried to 140 have the same volume, the difference being but 16 parts out of 100, the whole of the error cannot be very great, and one quarter six bushels of malt may, at a medium, be estimated of the same volume with one barrel of water. But, as experience is the surest guide, I have, from a very great number of different brewings, collected the following proportions, and repeatedly found them to be true. I have added, in the table, the weight malt ought to have, at every degree of dryness.

A TABLE *shewing the quantity of malt of every degree of dryness, equal to the volume of one barrel of water, and of the mean weight of one quarter in proportion to its dryness.*

	Degree of dryness.	Excise gauge without the bills. Volume of grain.	Excise gauge within the bills. Volume of grain.	Weight in pounds
Barley	80	1,56	1,59	376
	100	1,62	1,63	306
	105	1,62	1,67	301
	110	1,65	1,71	296
	115	1,67	1,75	291

Malt	119	1,68	1,79	286
	124	1,71	1,83	281
	129	1,74	1,87	276
	134	1,77	1,91	271
	138	1,80	1,95	266
	143	1,83	2,00	261
	148	1,86	2,03	256
	152	1,89	2,07	251
	157	1,92	2,11	246
	162	1,95	2,15	241
	167	1,98	2,19	236
	171	2,01	2,23	231
	176	2,04	2,27	226

With a table thus constructed, it is very easy to reduce every grist to its proper volume of water. Suppose those of the brewings we have already mentioned; that of the small beer consists of 6 quarters of malt dried to 130 degrees, the proportion of which in the table is as 1,75 to 1.

Quarter of malt. Barrel of water. Malt. Water.

If 1,75   1   6   3,42.

These six quarters of malt occupy therefore an equal volume with 3,42 barrels of water. A brown beer grist of 11 quarters dried to 130 degrees; the proportion of this in the table is as 1,74 to 1.

Malt.   Water. Malt. Water.

If 1,74   1   11   6,32

The volume of these 11 quarters of malt is therefore the same with that of 6,32 barrels of water, and the whole being brought to one denomination, we are enabled to find the heat of the first mash; but the effervescence occasioned by the union of the malt and water must prevent this calculation being strictly true, the consideration of which shall take place hereafter.

The circumstances are different in the other mashes: the waters used for these, meet a grist already saturated, and the volume is increased beyond the quantity found for dry malt. The quantity to be allowed for this increase cannot be determined by our former calculations, and new trials are to be made, in order to fix upon the true proportion.

Gauging is undoubtedly the most certain method of proceeding in these researches; but even this becomes less sure, on account of the expansion, evaporation, effervescence, and other incidents already mentioned.—Our errors however cannot be very considerable, when we deduce our conclusions from numerous and sufficiently varied experiments.

The volume of the grist of pale malt was found, after the parting of the first extract, to be 15,41 inches, though the space occupied by the malt, when dry, was only 5,51 inches: and the volume of the brown grist, at the same period, was 22,36 inches, though the dry malt filled only a space of 8,21 inches. The proportion in both these cases, and in all those which I have tried, answers nearly to one third, so that the volume of the grist, in the second and all subsequent mashes, may be estimated at three times the bulk of the malt when dry, and this is sufficiently accurate for the operations of brewing, in which, for conveniency sake, the application of whole numbers should be effected.

As it is found, by the gauges, that the goods, after the several taps are spent, remain sensibly of the same volume, or at least very little diminished; may we not conclude, the parts absorbed by the water, in which the virtue of the grain and the strength of the beer consist, are contained in an amazing small compass? It is indeed true that hot waters and repeated mashes do swell somewhat the hulls and skins of the malt, but no allowance made for this increase will be sufficient, to remove the cause of our surprise.

SECTION XI.

Of the Proportion of cold Water to be added to that which is on the point of Boiling, in Order to obtain the desired heat in the Extract.

THE degree of heat, which causes water to boil is determined, by Farenheit's scale, to 212. It is in our power to give to any part of the extracting water this degree of heat; and by adding to it a sufficient proportion of water of an equal heat with that of the air, and blending these two quantities with the grist, to bring the whole to the required temperature. The rules for obtaining this end are extremely simple, and cannot be unknown to those, who are skilled in arithmetical operations. But as our view is to render this part of our work generally useful, we think it will be proper briefly to lay down these rules, and to illustrate them by the examples of our two brewings.

Rule to ascertain the heat of the first Mash.

Let a express the degree of boiling water, b the actual heat of the air, c the required degree for the extract, m the whole quantity of water to be used, n the volume of the malt; x, that part of the water, which is to be made to boil, will be determined by the following equation.

$$x = \frac{\overline{c-b} \times \overline{m+n}}{a-b}$$

The quantity of water used, added (+) to the volume of the grist.

Their sum (z) multiplied (×) by the heat required, less (-) the heat of the air.

This produce divided (÷) by the heat of boiling water (212) less (-) the heat of the air will quote how much is to be made to boil or brought through (212) that is how high the copper is to be charged, the remainder of the length of the whole liquor for this mash, is the quantity to be cooled in.

The first example is that of a brewing of small beer, when the heat of the air is at 60, (see page 232.) The volume of the 6 quarters of malt was estimated at 3,42 barrels, (see page 268;) the first liquor is 14½ barrels, (see page 247) and the heat required for the first mash 154 degrees, (see page 247.)

First Mash.

$$m = \quad 14{,}50 \text{ Barrels of water}$$
$$n = \quad 3{,}42 \text{ Volume of grist}$$

$$m + n = \quad 17{,}92 \quad c = 154 \text{ Heat of the first mash,}$$
$$94 \quad b = 60 \text{ Heat of the air,}$$

(a) heat of $\qquad c - b = 94$

boiling water, 212   7168

b heat of the air, 60   16128

$a - b = 152 \,)\, 168448$

152

164 (1108 barrels of water, to be made to boil out of the 14 + 1/2 barrels which are allotted for the first mash. The incidents to be mentioned, are not considered in this calculation.

152

1248

1216

The next example of a brewing is that of a grist of eleven quarters of malt for porter or brown beer; the medium heat of the air is forty degrees, the volume of the grist, 6,32 barrels, (see page 268) the first liquor to mash with sixteen barrels, (see page 245) and the heat expected in the mash, one hundred and fifty-seven and a half[29] degrees. (See page 245).

First Mash of brown strong beer.

        16,00 Barrels of water

        6,32 Volume of malt

        ———   157 Heat required in the

        22,32   mash, vide page 247.

        117   40 Heat of the air.

        ——— ———

Heat of boiling   15624 117

   water,   212   2232 ———

Heat of air, 40 2232

        ——— ———

        172)261144 (15,18 barrels of water, to be

             172   made to boil out of the

           ——— 16 barrels.

           891

           860

           ———

           314

          1324

           ———

I will give one proof of the certainty of this rule, by setting down the state of this first mash from it.

        15,18

        212

       ———

	3036	
	1518	
	3036	
	———	
A.	3218,16	Number of degrees of heat in 14,66 barrels of boiling water.
	16,00	Barrels of water to first mash.
	15,18	Barrels made to boil.
	———	
	,82	Barrel to cool in.
	40	Heat of cold water.
	40	Heat of cold water.
B.	32,80	Number of degrees of heat in 1,34 barrels of cold water.
	15,18	Boiling water.
	,82	Cold water.
	6,32	Volume of grist.
	———	
C.	22,32	Barrels, volume of the whole mash.
	6,32	Barrels, volume of the 11 quarters of malt.
	,40	Heat of the grist.
	———	
	252,80	Number of degrees of heat in the grist.
	32,80	B.
	3218,16	A.
	———	

C. 22,32)350376
       2232
       (157 degrees of heat required in the first mash, as above.
———

12717

11160
———

15576

15624
———

So long as the mixture consists only of two quantities of different heat, as is always the case of the first mash, the preceding solution takes place. But in the second and other mashes, where three bodies are concerned, each of different heat, viz. the boiling water, the cold water, and the mash, are to be mixed, and brought to a determinate degree, the rule must be different; yet, like the former, it is the same with what is used in similar cases of allaying, when different metals are to be melted down into a compound of a certain standard, or different ingredients of different value to be blended, in order to make a mixture of a determinate price. What the different density of the metals, or the different value of the ingredients are, in these cases, the different degrees of heat of the boiling water, the grist, and the air, are in this.

Rule to ascertain the heat of the second mash, and of the subsequent ones.

Let the same letters stand for the things they signified before, and d express the actual heat of the grist, then will

$$x = \frac{\overline{c-b} \times m + \overline{c-d} \times n}{a-b}$$

or in plain terms, the heat required less (-) the heat of the air, multiplied (×) by the quantity of water used.

The heat required less (-) the heat of the goods, multiplied (×) by the volume of the goods.

Their sum (z) divided (÷) by the heat of boiling water, (212) less (-) the heat of the air.

Will quote the quantity to be made to boil, or to be brought through (212) the remainder part of the whole liquor for the mash is consequently the quantity to be cooled in.

We may now collect the circumstances of the two brewings, and find the quantity of boiling water, required for their second and subsequent mashes, exclusively of the incidents which will hereafter be mentioned.

The first mash for the six quarters of small beer, had 154 degrees of heat, but this and every mash loses, in the time the extract is parting from it, 4 degrees, which reduces the heat to 150 degrees. The volume of this grist, in its dry state, was 3,42 barrels, but now, by being expanded, and having imbibed much water, it occupies three times that space, or 10,26 barrels; the air is supposed to continue in the same state of 60 degrees of heat. The length and heat to be given to the three remaining mashes, are as follows. (See page 247.)

Degrees of heat,	154	166	170	174
Barrels of water,	14½	14½	11	11
Liquors,	1st	2d	3d	4th

        1 wort.   2 wort.

Second Mash for Small Beer.

$c =$ 166 Heat required in the mash.

$d =$ 150 Heat of the goods.

$c - d =$ 16

$n =$ 1026 Volume of the goods.

96

32

$$ 160$$

$$c - d \times n = \ 16416$$

$$c = \ 166 \text{ Heat required in the mash.}$$
$$b = \ 60 \text{ Heat of the air.}$$

$$c - b = \ 106$$
$$m = \ 1450 \text{ Barrels of water.}$$

$$ 5300$$
$$ 424$$
$$ 106$$

$$c - b \times m = 153700$$
$$c - d \times n = \ 16416$$

$a - b = 152$) 170116 ($11,19$ Barrels of water to be made to
$ 152$ boil out of the quantity allotted
$ \overline{}$ for the second mash.
$a = 212 181$
$b = 60 152$
$ \overline{} \overline{}$
$ 152$
$ 291$
$ 152$
$ \overline{}$
$ 1396$
$ 1368$

Third Mash.

170 Heat of mash.	170 Heat of 3rd mash.
60 Heat of air.	162 Heat of goods.
———	———
110	8
1100 Barrels of water	1026 Volume of grist.
———3d mash.	———
11000	8208
110	
———	
121000	
8208	
———	

152)129208

1216 (8,50 Barrels to be made to boil out of the quantity of water allowed for the third mash.

———

760

760

———

8

Fourth Mash.

174 Heat of 4th mash.	174 Heat of 4th mash.
60 Heat of air.	166 Heat of goods.
———	———
114	8
11,00 Barrels of water	1026 Volume of goods.
———for 4th mash.	———

```
        11400                              48
          114                              16
         ____                              80
       125400                             ____
         8208                            8208
         ____

    152)133608
         ____
         1216 (879 Barrels to be made to boil out of the quantity of water
              allowed for the fourth mash.
         ____
         1200
         1064
         ____
         1368
         1368
         ____
```

The liquors of this brewing of common small beer, when the mean heat of the air is 60 degrees, must therefore be ordered in the following manner (the incidents hereafter to be noticed, excepted.)

	1 Liqr.	2 Liqr.	3 Liqr.	4 Liqr.
Lengths of liquors,	14½	14½	11	11
Boiling water; barrels,	11	11½	8½	8¾
Cold water; barrels,	3½	3¼	2½	2¼
	14½	14¼	11	11

The heat of the first mash for the 11 quarters of brown beer, was 157 degrees, (see page 245) and after the parting of the extract from it, 153; the volume of the grist, in its dry state, was valued at 6,32 barrels of water, (see page 268) but, for the reasons before mentioned, it now occupies three times that space, or 18,96 barrels. The air is supposed to continue at 40 degrees, and the length and heat to be given to the different mashes, were determined as follows: (see page 245.)

Degree of heat,	157	158	162	164	165
Barrels of water,	16	8	12	9	9
Liquors;	1st	2d	3d	4th	5th

1 wort.　　2 wort.　　3 wort.

Second Mash of Porter, or brown strong.

212 Boiling water.

40 Heat of air.

———

72

　　　　　　　　　　　　　　158 Heat of 2nd mash.

　　　　　　　　　　　　　　158 Heat of the grist or goods.

　　　　　　　　　　　　　　———

　　　　　　　　　　　　　　———

　　　　　　　　　　　　　　5

　　　　　　　　　　　　　　1896 Volume of goods.

　　　　　　　　　　　　　　———

　　　　　　　　　　　　　　———

158 Heat of 2nd mash　　　　30

40 Heat of air　　　　　　　　45

———　　　　　　　　　　　　40

118	5
8,00 Barrels of ———— water.	——— ——— 9480 ——— ———
94400	
9480	
———	

172)103880 (6,03 Barrels of water to be made to
  1032 boil for the second mash.
  ———
  680
  516
  ———

Third Mash.

212 Heat of boiling water.
 40 Heat of air.
——— Heat of air.
172

      162 Heat of 3rd mash.
      154 Heat of goods.
      ———

162 Heat of 3rd mash.	8
40 Heat of air.	18,96
———	———
122	48
12,00 Bar. of water.	72

146400	64
15168	8
——	——
172)161568	15168
1548	(9,45 Barrels of water to be made to boil for third mash.
776	
688	
888	
860	

Fourth Mash.

164 Heat of 4th mash.

158 Heat of goods.

——

6

18,96 Volume of grist wetted.

164 Heat of 4th mash.	——
40 Heat of air.	36
——	54
124	48
9,00 Bars. of water.	6
——	——
111600	11376

11376

172)122976 (7,14 Barrels of water to be made to boil for the fourth mash.
1204

257
172

856
688

168

Fifth Mash.

165 Heat of 5th mash.
160 Heat of Goods.

5
18,96

165 Heat of 5th mash.
40 Heat of air.

30
45

125
40

9,00 Barrels of water.
5

114500
9480

9480

172)123980 (7,20 Barrels of water to be made to boil for the 5th mash.
　　1204

　　　358
　　　344

　　　140

The liquors of this brewing of brown beer must therefore be ordered in the following manner:

Barrels of boiling water,	15¼	6	9½	7	7
Barrels of cold water,	¾	2	2½	2	2
	16	8	12	9	9
Liquors,	1st.	2nd.	3rd.	4th.	5th.

What in the brewery is generally called cooling in, must be settled for this brewing according to the number of barrels of cold water specified as above, the incidents hereafter to be noticed excepted.

Each of these calculations may be proved in the same manner as was done before. This method of discovering the proportion of water to be cooled in, deserves, on account of its plainness and utility, to be preferred to any other, which depend only upon the uncertain determination of our senses.

SECTION XII.

OF MASHING.

OF late years, great progress has been made towards perfecting the construction and disposition of brew-house utensils, which seem to admit of very little farther improvement. The great copper, in which the waters for two of the extracts receive their temperature, is built very near the mash tun, so that the liquid may readily be conveyed to the ground malt, without losing any considerable heat. A cock is placed at the bottom of the copper, which being opened, lets the water have its course, through a trunk, to the real bottom of the mash tun. It soon fills the vacant space, forces itself a passage through many holes made in a false bottom, which supports the grist, and, as the water increases in quantity, it buoys up the whole body of the corn.

In order to blend together the water and the malt, rakes are first employed. By their horizontal motion, less violent than that of mashing, the finest parts of the flower are wetted, and prevented from being scattered about, or lost in the air.

But as a more intimate penetration and mixture are necessary, oars are afterwards made use of. They move nearly perpendicularly, and by their beating, or mashing, the grains of the malt are bruised, and a thorough imbibition of the water procured.

The time employed in this operation cannot be settled with an absolute precision. It ought to be continued, till the malt is sufficiently incorporated with the water, but not so long as till the heat necessary to the grist be lessened. As bodies cool more or less speedily, in proportion to their volume, and the cohesion of their parts, a mash which has but little water, commonly called a *stiff mash*, requires a longer mashing to be sufficiently divided, and, from its tenacity, is less liable to lose its heat. This accounts for the general rule, that the first mash ought always to be the longest.

After mashing, the malt and water are suffered to stand together unmoved, generally for a space of time equal to that they were mashed in. Was the extract drawn from the grain as soon as the mashing is over, many of the particles of the malt would be brought away undissolved, and the liquor be turbid, though not rich. But, by leaving it some time in contact with the grain, without any external motion, many advantages are gained. The different parts of the extract acquire an uniform heat, the heaviest and most terrestrial subside, the pores being opened, by heat, imbibe more readily the water, and give way to the attenuation and dissolution of the oils. When the tap comes to be set, or the extract to be drawn from the grist, as the bottom of the mash is become more compact, the liquor is a longer time in its passage through it, is in a manner strained, and consequently extracts more strength from the malt, and becomes more homogeneous and transparent.

Such are the reasons why the grist should not only be mashed pretty long, but likewise be suffered to rest an equal time. It is the practice of most brewers, and experience shews it is best, to rake the first mash half an hour, to mash it one hour more, and to suffer it to stand one hour and a half. The next extract is commonly mashed three quarters of an hour, and stands the same space of time; the third, and all that follow, are allowed one half hour each, both for mashing and standing.

The heat of the grist being in this manner equally spread, and the infusion, having received all the strength from the malt, which such a heat could give it, after every mashing and standing, is let out of the tun. This, undoubtedly, is the fittest time to observe whether our expectations have been answered. The thermometer is the only instrument proper for this purpose, and ought to be placed, or held, where the tap is set, adjoining to the mouth of the underback cock. The observation is best made, when the extract has run nearly half; and as, by it, we are to judge with what success the process is carried on, it is necessary to examine every incident, which may cause a deviation from the calculated heat.

SECTION XIII.

Of the Incidents, which cause the Heat of the Extract to vary from the Calculation, the allowances they require, and the means to obviate their effects.

BY incidents, I understand such causes as effect either the malt, the water, or the mash, during the time the brewing is carrying on, so as to occasion their heat to differ from what is determined by calculation. As these might frequently be a reason of disappointment, an inquiry into their number and effects will not only furnish means to prevent and rectify the errors they occasion, but also serve to confirm this practice.

In our researches on the volume of malt, some notice was taken of the increase of bodies by heat, and the loss occasioned by evaporation. Water, when on the point of ebullition, occupies the largest space it is susceptible of; but contracting again, when cold water is added to it, the true volume of both, when mixed together, remains uncertain, and may cause a difference between the calculated and real degree of heat. This cause, however, producing an effect opposite to, and balanced in part by evaporation, becomes so inconsiderable, as hardly to deserve any farther consideration.

Water, just on the point of ebullition, may be esteemed heated to 212 degrees. Though, by the continuation of the fire, or by any other cause[30], the heat never goes beyond this, yet was cold water added to that, which violently boils, the degree expected from the mixture would be exceeded; for the cold water absorbing the superfluous quantity of fire, which otherwise flies off, becomes hot itself, and frustrates the intent. The time, therefore, of adding the cold water to the hot is immediately before the ebullition begins, or when it is just ended; and in proportion as we deviate from this practice, the heat in the extract will differ from the calculated degree.

The water, for every mash, should, as near as possible, be got ready to boil, and be cooled in just before it is to be used. A liquor, which remains a long time after the ebullition is over, and the fire has been damped up, loses part of its heat, if cold water is applied to it, the effect cannot be the same as it would have been at first. On the contrary, if the liquor is got ready too soon, and cold water immediately added to it, in order to gain the proper degree of temperature, by leaving the mixture long together, though the fire is stopped up, more heat than necessary will be received from the copper and brickwork, especially if the utensils are large. In both cases, the degree in the extract will not answer the intent.

The effect of effervescence next deserves our consideration, but this takes place only when the water first comes in contact with the malt. Germinated grains must, to become malt, be dried so, that their particles are made to recede from one another, thus deprived of the parts, to which their union was due, when they come in contact with other bodies, (as water) they strongly attract the unitive particles they want, and excite an intestine motion, which generates heat. This motion and this heat are more active in proportion as the grain has more strongly been impressed by fire, and the extracting water is hotter.

A large quantity of liquor applied to the grist is less heated than a small one, by the power of effervescence. The least quantity of water, necessary to shew that power, must be just so much as the malt requires to be saturated, which we have seen to be double the volume of the grain. When more water than this is applied to the grist, the real effervescing heat is by so much lessened, being dispersed in more than a sufficient space.

A table shewing the heat of effervescence for every degree of dryness in the malt, can only be formed from observations. To apply this table to practice, and to find out, for any quantity of water used in the first mash, the degrees of heat produced by effervescence, three times the volume of the grist must be multiplied by the number expressing the effervescing heat for malt of such a degree of dryness, and this produce be divided by the real volume of the whole mash.

A TABLE *shewing the heat occasioned by the effervescing of malt, for its several degrees of dryness.*

Dryness of malt.	Heat of effervescence.
119°	0
124	3½
129	7
134	10½
138	14
143	17½
148	21
152	24½
157	28
162	31

167	35
171	38½
176	40

Malt dried only to 119 degrees, raises no effervescence, and the strongest is generated by malt dried to 176 degrees; the heat produced by this amounts to 40 degrees, but the number of effervescing degrees, in this or any other case, are reached but from success attending our endeavours, ultimately to penetrate the malt by heated water, or not until the grist is perfectly saturated, which, in point of time, generally takes up the whole space of the first mashing and standing; the air, therefore, cannot cause any diminution of heat, an incident which affects considerably every subsequent mash.

The little copper being more distant from the mash tun than the other, the water there prepared, in its passage to the goods, loses some part of its heat. And in proportion to the quantity of water used, to the number of the extracts that have been made, and according as the mashes have more or less consistency, in the same time do they part with more or less of their heat. Observations made separately upon strong and small beer, have shewn the proportions of this loss to be as follows:

For strong beer.

Mashes 2d 3d 4th 5th

Heat lost 8° 12° 8° 8°

For small beer.

Mashes 2d 3d 4th

Heat lost 8° 16° 20°

A grist not perfectly malted, or one which contains many hard corns, disappoints the expectation of the computed degree, as the volume cannot be such as was estimated from an equal dryness of true germinated grain. It has been observed, that, in perfect malt, the shoot is very near pressing through the exterior skin. By so much as it is deficient in this particular, must it be accounted only as dried barley, or hard corn. I know no better way of judging what proportion of the corn is hard to what is malted, than by putting some in water, the grains not sufficiently grown will sink to the bottom. Were this to be done in a glass cylinder, the proportion between the hard and malted corn might be found with exactness.—The unmalted parts being estimated with regard to their volume, as barley, a quarter of them will be to the barrel of water as 1,56 to 1[31]. Supposing, therefore, that, in the brown beer grist, before mentioned, the proportion of hard corns is of two quarters out of eleven, to discover the true volume of such a grist, the following rule may be used.

9 quarters of true malt	2 quarters of hard malt
1,74 volume at 130° of dryness	1,56 volume of
———	——— 1 quarter
15,66	3,12
3,12 volume of 2 quarters of hard corn	
Total ———	

numb. 11) 18,78 (1,70 true volume of one quarter of this malt to one barrel of water, and consequently the eleven quarters will fill a space equal to that of 6,47 barrels.

By means of this rule, we may find what increase of heat any proportion of hard corns will occasion, as will be seen in the following table.

Proportions of hard corns 1/4 1/6 1/8 1/16 1/32 of the grist

Greater heat of the mash   4°   3   2   1   1/2   degrees.

But the brewing of such malt ought to be avoided as much as possible, as the hard parts afford no strength to the extract.

If a grist is not well and thoroughly mashed, the heat not being uniformly distributed in the different parts of the extract, the liquor of the thermometer, when placed in the running stream of the tap, will fluctuate, and, at different times, shew different degrees of heat. In this case, the best way is to take the mean of several observations, and to estimate that to be the true heat of the mash.

If the gauges of the coppers are not exactly taken, a variation must be expected.

Though the small and hourly variations in the state of the atmosphere have but little influence upon our numbers, a difference will be observed in any considerable and sudden changes either of the heat or of the weight of the air. Our instruments, and in particular the thermometer, are supposed to be well constructed and graduated. If the water cooled in with is more or less hot than estimated, or if the time of mashing or standing is either more or less than was allowed for, the computation must be found to vary from the event.

While the malt is new, if the fire it has received from the kiln has not sufficiently spent itself, this additional heat is not easily accounted for. This is likewise the case, when malt is laid against the hot brickwork of coppers; and, on the contrary, a loss of dryness may be occasioned, if the store rooms are damp.

The artist should be attentive to all these incidents; the not pointing them out might appear neglectful; enumerating more would exceed the bounds of use.

Small grists brewed in large utensils lose their heats more readily, by laying thin, and greatly exposed to the air; and, on the contrary, a less allowance, for the loss of heat, is required in large grists, and to which the utensils are in proportion.

This really is the only difference between brewings carried on in large public brewhouses, and those made in small private places, in other respects constructed upon the same plan, and with an equal care. Prejudice has propagated an idea, that where the grists are large, and the utensils in proportion, stronger extracts could be forced from the malt, in proportion to the quantity, and that more delicate beers could be made in smaller vessels less frequently used. These assertions, from what has been said, will, I hope, need no farther enquiry: the degrees of heat for the extracts are fixed for every intent, and it cannot be advantageous, by any means, to deviate from them. Brewings will most probably succeed in all places, where the grist is not so large as to exceed the bounds of man's labour, and not so small as to prevent the heat from being uniformly maintained. The disadvantages are great on all sides, when a due proportion is not observed between the utensils and the works carried on.

It will now be proper to continue the delineation of our two brewings, and to put all the circumstances relating to them under one point of view.

A brewing for porter or brown strong beer, computed for 40 degrees of heat in the air.

11 quarters of malt, dried to 130 degrees, 132 pounds of hops for 27 barrels 1/2, to go out at 3 worts, 31 Inches above brass.

```
         Volume of grist     6,32
         Water for first    16,00
         mash
                            ─────
                            22,32
```

```
  6,32 Volume of grist        6 effervescing degrees.
     3                        3 degrees for hard corns.
  ────                        ───
 18,96                        9
     7 Effervescence,           degrees equal to 2 inches 1/4 less
       per table.               cooling in for the first mash, (see
  ────                          page 152.)
22,32) 132,72
       13392   (6 degrees of heat gained in the first mash by effervescence.
```

Mashes	1st	2d	3d	4th	5th	
Deg. of heat,	157°	158°	162°	164°	165°	See p. 280.
Barrels of water used,	16	8	12	9	9	See p. 284.
Quantity ooled in by calculation,	3/4	2	2½	2	2	See p. 284.
Boiling water by calculation;	───	───	───	───	───	
barrels,	15¼	6	9½	7	7	
Allowances for incidents,	[32]G. C. }[33]Less 2 inches 1/4.	L. C. more 2 in.[34]	L. C. more 3 in.[34]	L. C. more 2 in.[34]		

A brewing for common small beer, computed for 60 degrees of heat in the air.

6 quarters of malt dried to 130 degrees; 36 pounds of hops;

30 barrels 3/4 to go out 56 inches above brass.

$$\begin{array}{lr} \text{Grist} & 3{,}42 \\ \text{Water} &)14{,}50 \\ \hline & 17{,}92 \end{array}$$

Volume of grist 3,42

3

———

10,26   4° for effervescence.

7 effervescing degree   1° for hard corns

for malt at 130   3° for new malt hot

——— (see table page 292.)   —

                                     8° to be deduced

17,92) 71,82 (4 degrees of heat   from the first

       7168 gained in the mash   cooling in.

       ——— by effervescence.

        14

Mashes	1st	2d	3d	4th	
Deg. of heat.	154	166	170	174	See p. 248.
Whole quantity of water used, barrels	14½	14½	11	11	See p. 280.
Quantity to be cooled in, barrels	3½	3¼	2½	2¼	See p. 280.

Boiling water by calculation		—	—	—	—	
charged, barrels		11	11¼	8½	8¾	See p. 280.
Allowances for incidents;		G. C.[35]	G. C.[36]	L. C.[36]	L. C.[36]	
inches;		less 2;	more 2;	more 4;	more 5;	

These computations, perhaps, will appear more troublesome than they really are; but, besides the facility which exercise always gives for operations of this kind, the satisfaction of proceeding upon known principles, will, I hope, encourage the practitioner to prefer certitude to doubt. One advantage must greatly recommend it, and at the same time secure the uniformity of our malt liquors; tables for each sort and season may be made beforehand, and will serve as often as the circumstances are the same. The trouble of the computations will by that means be saved, and by collecting together different brewings of the same kind, the artist will, at any time, have it in his power to see what effect the least deviation from his rules had upon his operations, and to what degree of precision he may hope to arrive.

That nothing may be wanting in this work, to facilitate the intelligence thereof, I shall insert the method of keeping the account of actual brewings, made according to the computations I have here successively traced down. The first column contains the charges of the coppers, and the numbers computed; the next, the brewings made from these numbers, with their dates, and the degrees of heat found by observation; the variations occasioned by unforeseen incidents are supposed to be allowed for, at cooling in, by the artist, upon the principle, that each inch of cooling in answers to four degrees of heat. Noting in this manner the elements of every brewing we make, when the drink comes into a fit state for use, we are enabled to compare our practice with the principles which directed it; by this means, experiments constantly before our eyes will be the most certain and best foundation for improvement.

Small Beer. Heat of air 60 Degrees. 6 quarters of Malt, 36lb. of Hops, for 30 Barrels 3/4, to go out 56 Inches above Brass.

Observations.

1760. 1760. 1760. 1761. 1761.
June. 27. June. 30. Aug. 3. July. 3. July. 3.

1st Liquor. Charge great copper, 2 inches 1/2 above brass; cool in to 13 inches 1/2 above brass, rake 1/2 hour, mash 1 hour, stand 1 hour 1/2, heat of the extract intended 150 degrees. } 151   149   150   153   150

2d Liquor. Charge great copper 7 inches above brass, cool in to 13 inches 1/2 above brass, mash 3/4 hour, stand 3/4 hour, heat intended 162 degrees—1 wort came in 33 inches above brass, boiled 1 hour 1/2, went out 28 inches above brass. in to 13 inches 1/2 above brass, } 161   163   163   148   162

3rd Liquor. Charge little copper 3 inches 1/2 above brass, cool in to 13 inches above brass, mash 1/2 hour, stand 1/2 hour, heat expected 166 degrees—2 } 166   165   165   167   15

4d Liquor. Charge little copper 10 inches 1/2 above brass, cool in to 13 inches above brass, mash 1/2 hour, stand 1/2 hour, heat expected 170 degrees—2 wort came in 39 inches above brass, boiled down to 28 inches above brass. } 166   165   165   167   15

    Length 31 barrels.

Porter. *Heat of the Air 40 Degrees. 11 quarters of Malt, 132lb. of Hops for 27 Barrels 1/2, to go out at 3 Worts, 31 Inches above Brass.*

                                    1761. 1761. 1761.
                                Nov. 20. Nov. 3. Nov. 3.
                                      43°   35°   40°

1st Liquor. Charge great copper 13 inches above brass, cool in to 17 inches 1/2 above brass, rake 1/2 hour, mash 1 hour, stand 1 hour 1/2, extract expected 153 degrees. } 153 151 154

2d Liquor. Charge little copper 2 inches 1/2 below brass, cool in to 3 inches above brass, mash 3/4 hour, stand 3/4 hour, heat expected at the tap 154 degrees—1 wort came in great copper 16 inches above brass, boiled 1 hour, went out 13 inches above brass. } 155 157 153

3d Liquor. Charge little copper 10 inches above brass, cool in to 16 inches above brass, mash 1/2 hour, stand 1/2 hour, the tap to come down 158 degrees—2 wort came in great copper 11 inches above brass, boiled 2 hours, went out 5 inches above brass. } 157 157 158

4th Liquor. Charge little copper 1 inch 1/2 above brass, cool in to 6 inches 1/4 above brass, mash 1/2 hour, stand 1/2 hour, to come down 160 degrees. } 161 160 161

5th Liquor. Charge little copper 1 inch 1/2 above brass, cool in to 6 inches 1/4 above brass, mash 1/2 hour, stand 1/2 hour, tap to come down at161 degrees—3 wort came in 25inches above brass, went out 13 inches above brass. } 160 161 160

SECTION XIV.

Of the disposition of the Worts when turned out of the Copper, the thickness they should be laid at in the Backs to cool, and the heat they should retain for fermentation, under the several circumstances.

WHEN a process of brewing is regularly carried on with two coppers, the worts come in course to boil, as the extracts which formed them are produced. It would be tedious and unnecessary to describe the minutest parts of the practice; which, in some small degree, varies as brewing offices are differently constructed, or the utensils are differently arranged. Without the assistance of a brewhouse, it is perhaps impossible to convey to the imagination the entire application of the rules before laid down, but with one, I hope they need little, if any, farther explanation.

The worts, when boiled, are musts possessing an intended proportion of all the fermentable principles, except air; this was expelled by fire, and until their too great heat is removed, cannot be administered to them.

In musts, which spontaneously ferment, the external air excites in their oils an agitation, which, heating and opening the pores of the liquor, expands and puts in action the internal air they possess. The case is not exactly the same with regard to those musts which require ferments. The air wanted in boiled worts must be supplied by the means of yeast. Was the heat of the wort such, as to occasion the immediate bursting of all the air bubbles contained in the yeast, an effervescence rather than a fermentation would ensue. Now a heat superior to 80 degrees has this effect, and is therefore one of the boundaries in artificial fermentation; 40 degrees of heat, for want of being sufficient to free the air inclosed in the yeast bubbles, and to excite their action, is the other. Within these limits, must the wort be cooled to; and the precise degree, which varies according to the different circumstances they are in, and to the intent they are to be applied to, is, together with the means of procuring this heat, the purport of this section.

Worts, when in the copper, boil at a heat somewhat superior to that of 212 degrees; the more this is exceeded, the stronger the liquor is. The instant the wort is suffered to go out of the copper, it loses more heat than in any other equal space of time after it has been exposed to the air. In the course of the natural day, or in 24 hours, the heat of the air varies sometimes, (especially in summer) as much as 20 degrees. If the wort, after having reached the lowest heat in this interval, was suffered to remain in the coolers, till the return of a greater in the air, it would be influenced by this increase, expand, and be put in action; and, should there be at this time any elastic air in any part of the coolers, which sometimes happens, either from the sediment of former worts, from the backs not being clean swept, or from the wood being old and spungy, the wort supposed to be left to cool, will, by receiving the additional heat from the air, and blending with the incidental elastic air adhering to the coolers, bring on, in a lower degree, the act of fermentation; an accident by the artist called the *backs being set*.

For this reason, a wort should never be suffered to lay so long as to be exposed to the hazard of this injury, which generally may happen in somewhat more than twelve hours. Thus are we directed to spread or lay our worts so thin in the backs, as they may come to their due temperature within this space; in summer it is sufficient if the backs be covered; in winter a depth of two inches may oftentimes be allowed with safety.

From the inclination of the coolers or backs to the place, where the worts run off, from their largeness, or from the wind and air warping them, a wort seldom, perhaps never, lays every where at an equal depth, and cannot therefore become uniformly cold in the same space of time. This renders the use of the thermometer difficult, though not impracticable. To supply the want of this instrument with some degree of certainty, the hand intended to feel the worts, is brought to the heat of the body, by placing it in the bosom, until it has fully received it. Then dipping the fingers into the liquor, we judge, by the sensation it occasions, whether it is come to a proper degree of coolness to be fermented. As the external parts of our bodies are generally of about 90 degrees of heat, some degree of cold must be felt, before the worts are ready for the purpose of fermentation. But that degree varies for different drinks, and in different seasons. I will endeavour to point out the rules to form a judgment for the heat of small beer worts. A greater precision, both for that and for other drinks, will be found in the following table.

In July and August, no other rule can be given, than that the worts be got as cold as possible. The same rule holds good in June and September, except the season is unnaturally cold. In May and October, worts should be let down nearly thirty degrees colder than the hand; in April, November, and March, the worts should be about twenty degrees colder than the hand, and only ten in January, February and December.

It may perhaps be thought that the heats here specified are great, but worts cool as they run from the backs to the working tuns, they are also affected by the coldness of the tuns themselves, and perhaps these circumstances are not so trivial, but that an allowance should be made for them. In general, the heat of no must should exceed 60 degrees, because fermentation increases this or any other degree, in proportion to that, under which this particular part of the process begins. To render the thermometer more useful, and to suit it to our conveniency, we have before supposed every first mash for common small beer to be made at four o'clock in the morning: in this case, and where the worts are not laid to cool at more than one inch in depth, the following table may be said to be a measure of time, the first and last worts for this drink should be let down at.

A TABLE, *shewing nearly the times the first and last worts of common small beers should be let down in the working tuns, supposing the first mash of the brewing to be made at four o'clock in the morning, and no uncommon change happens in the heat of the air.*

	Air	1st Wort.	2nd Wort.	
	30	3 o'clock.	5 o'clock.	
	35	3½	6½	
	40	4¾	8	C
	45	4¾	9¼	
	50	5½	11½	
A	55	6½	B 1½	
	60	7½	2½	
	65	8	2½	
	70	8½	3	D
	75	9	3	
	80	9	4	

A: Heat of the air at 8 o'clock in the morning.
B: Hours in the afternoon, same day as brewing began.
C: Hours in the afternoon.
D: Hours of the next morning.

Small beer worts being nearly alike in consistency, the necessary variations from this table must be less frequent. It is true, some difference may happen from the exposition of a brewhouse, or from other circumstances, admitting more or less freely the intercourse of the air, and be such as might alter, upon the whole, the times set down in the preceding page. Brown beer worts, which are more thick and glutinous, and especially amber worts, which are stronger still, will require other and longer terms to come to their due temperature, to be fermented at; but when once observed and noted, according to various degrees of heat in the air, at 8 o'clock each morning, the conveniency of these observations must be such, in this business, which requires long watchings and attendance, that no arguments are necessary to recommend what is rather indulgence than industry.

A TABLE *shewing the degrees of heat worts should be at, to be let down from the coolers into the working tuns, according to the several degrees of heat in the air.*

Heat of the air.	Common small.	All-keeping beers.	Amber or ales.
25	75	59	55
30	70	56	54
35	65	53	53
40	60	50	52
45	55	50	51
50	50	50	50
55 } 60	In these cases, when the medium heat of the air is greater than that which the worts should ferment bring them as near as possible to their temperature. It has been observed, that the coldest part of the natural day is about one hour before sun rising.		

The consequences of worts being set to ferment at, in an undue heat, are the following. In strong beers, or such as are intended for long keeping, if the worts be too cold, a longer time is required for their fermentation, and the drinks grow fine with more difficulty; if, on the contrary, they are too hot, acidity, and a waste of some of the spiritous parts must ensue. Either of these disadvantages appears more conspicuous in common small beer, as, in winter, this drink is seldom kept a sufficient time to correct the defect, and in summer, from being too hot, it becomes putrid, or, in the terms of the brewery, is hereby *foxed*.

SECTION XV.

Of Yeast, its nature, and contents, and of the manner and quantities in which it is to be added to the worts.

MUSTS, or worts, though ever so rich, when unfermented, yield no spirit by distillation, nor inebriate, if drank in any quantity. The oils, as yet not sufficiently attenuated for this purpose, become so only by fermentation. Air is absolutely necessary for this process, in the course of which, some of the aerial parts mixing with, and being enveloped by, oils greatly thinned, are enclosed in vesicles not sufficiently strong to resist the force of elasticity, or prevent a bursting and explosion. In the progress of the act, the air joins with oils both coarser, and charged with earthy particles, a coat is formed capable of resisting its expansion, and if the bubbles cannot come to a volume sufficient to be floated in and upon the liquor, they sink to the bottom, and take the appellation of *lees of wine*. Between these two extremes, there is another case, when the bubbles are sufficiently strong to hold the air, but not weighty enough to sink. After floating in, they emerge, and are buoyed upon the surface of the liquor, and there remaining entire, are termed the *flowers of wine*. Both lees and flowers are, therefore, vesicles formed out of the must, filled with elastic air, and, either separately, or when mixed together, they obtain the general denomination of *yeast*. We have often mentioned the power of fire, in driving the air out of worts. Yeast, fraught with the principle now wanted for fermentation, is, therefore, the properest subject to be added to the must; but its texture is various, in proportion to the different heats of the extracts it was formed from. Keeping drinks, extracted with hotter waters, yield yeast, the oils of which have a greater spissitude. It is consequently slower, more certain, and most fit to promote a cool and gentle fermentation. That, on the contrary, which is produced from small beer, being weak, and acting at once, is apt to excite a motion like that of effervescence; such yeast ought, therefore, not to be used, but when there is no possibility to obtain the other. The longer wines or beers are under the first act of fermentation, the greater variety will be found in the texture of the bubbles, which compose their flower and lees. Wines made out of grapes, in general, require a time somewhat longer than the worts of malt, before this first period is at an end; and we have seen, that in them fermentation first brings forth air bubbles, whose constituent parts are most tender, and afterwards some that are of a stronger texture. As malt liquors require a less time to ferment, their bubbles are more similar: on this account, the whole quantity of yeast necessary to a wort should not be applied at once, lest the air bladders, bursting nearly in the same time, should prevent that gradual action, which seems to be the aim of nature in all her operations. Keeping beers, formed from low dried malts, occasion the greatest variety of heat in the extracts, and from hence these musts form yeast, whose bubbles differ most in magnitude and strength. A drink, then, properly made from pale malt, nearly resembles natural wines, especially when they are so brewed as to require precipitation to become transparent. *Cleansing* is dividing the drink into several casks; this checks the motion occasioned by fermentation, and consequently retards it. To prevent this from being too sensibly felt, some yeast should be put to the drink, before it is removed into the casks. As the constituent parts, in strong beers, are more tenacious than in small, and require a greater motion to entertain the fermentation, the drinks, before they be thus divided, should, besides the addition of the yeast, be well roused with a scoop, or by some other means, for one hour. This not only blends all the parts together, but attenuates and heats the liquor, and makes it more ready to begin to ferment again, when in the casks. One sixth part of the whole of the yeast used is generally reserved for this purpose; and the remainder is equally divided as the worts are let down. It must be observed, that this stirring, though as necessary to small, as to strong drinks, is only to be continued for a space of time proportioned to their strength. We have before seen, when a grist of malt is entirely extracted to form common small beer, soon to be expended, one gallon of yeast to eight bushels of grain affords a sufficient supply of air to perfect the fermentation. This takes place when the heat of the air is at 40 degrees, but, at the highest fermentable degree, experience shews, that half that quantity is as much as is necessary. For some ales, the whole virtue of the malt is not extracted, and what remains is appropriated to the making of small beer: the quantity of yeast used for these drinks must be only in proportion to the strength extracted. From these premises, the following tables have been formed, exhibiting the quantity of yeast proper for the several sorts of drinks, at the different heats of the air.

A TABLE *shewing the quantities of yeast necessary for common small beer in every season.*

Heat of the air. Pints of yeast to one quarter of malt.

Heat	Pints	Notes
35	9	
40	8	} The whole quantity of yeast to be put into the first wort.
45	8	
50	7	———————————————————————
55	7	} The first wort to have ¾
60	6	} The second wort to have ¼
65	6	———————————————————————
70	5	}

75	5 The first worth to have one half of the whole quantity.
80	4 The second worth to have the remainder.

A TABLE *shewing the quantities of yeast necessary for all keeping drinks, both brown and pale, small and strong.*

Heat of the worts[37] Pints of yeast to one quarter of malt.

30	6
35	6
40	6
45	6
50	6
55	5
60	5[38]
65	4
70	4
75	3
80	3

A TABLE *shewing the quantities of yeast necessary for amber and all sorts of ales, after which small beer is made.*

Heat of the air. Pints of yeast to one quarter of malt.

30	7½
35	7
40	7
45	6½
50	6
55	5½
60	5

65	4½
70	4
75	3½
80	3

This table is founded on the supposition that, the virtue or strength extracted from one quarter of malt for amber, is equal to 5/6 of the whole. In every heat of the air, the quantity of yeast to be used for common small beer made after ale, must be one fifth part of the quantity which the ale required, the additional strength obtained from reboiling the hops, requiring further proportion; if, for keeping small beer, nearly in the proportion of six pints of yeast to five barrels of beer, this will be found to correspond with the rule delivered in the foregoing table.

SECTION XVI.

Of practical Fermentation, and the management of the several sorts of Malt Liquors, to the period, at which they are to be cleansed or put into the Casks.

THE laws of fermentation are universal and uniform; and when it proceeds regularly, its different periods are known by the different appearances of the fermenting liquor. As a particular appellation is given to each of these, it may not be unnecessary here to describe them.

1. The first sign of a wort fermenting is a fine white line, composed of very small air bubbles, attached to the sides of the tun; the wort is then said to *have taken yeast*.

2. When these air bubbles are extended over the whole surface of the must, it is said to be *creamed over*.

3. Bubbles continuing to rise, a thin crust is formed; but as the fermentation advances rather faster near the sides of the tun, than in the middle, this crust is continually repelled; from which arises the denomination of *the wort parting from the tun side*.

4. When the surface becomes uneven, as if it were rock work, this stage of fermentation, which has no particular use, is distinguished by its *height*.

5. When the head becomes lighter, more open, more uniform, and of a greater depth, being round or higher in the middle, than in any other part, and seeming to have a tendency still to rise, the liquor is denominated to be of *so many inches, head not fit to cleanse*.

6. This head having risen to its greatest height, begins to sink, to become hollow in the middle, and, at the same time, more solid, the colours changing to a stronger yellow or brown; the wort is then said to be *fit to cleanse*.

After this, no farther distinctions are made; if the fermentation is suffered to proceed in the tun, the head continues to sink, and the liquor is often injured.

As the denominations and tastes of liquors brewed from malt are numerous, it is impossible to specify each separate one; we shall therefore only particularize such sorts of drinks, as were taken notice of in the section of extraction, they being most in use; but, from what will be said concerning them, the method of managing any other malt liquor may easily be deduced.

Spontaneous pellucidity arises from a due proportion of the oils to the salts, in the worts, but the advantage of long keeping depends not only on the quantity of oils and hops the musts possess, but also on the fermentation being carried on in a slow and cool manner. All drinks, intended long to be kept, are therefore best formed in cold weather, and made to receive their yeast at such temperature, as is set forth in the table. The yeast is to be divided in proportion to the quantities of wort let down, until the whole, being mixed together, receives its allotted portion, except so much as is to be put in just before cleansing. Under these circumstances, drinks, which are brewed for keeping, are suffered to go through the first process of fermentation, till they are so attenuated, that the liquor becomes light, and the head, or the yeast, laying on the surface of the beer, begins to sink. When, or somewhat before, this head has fallen to nearly half the greatest height it reached to, a remarkable vinous smell is perceived, and the liquor, at this term, is to be put into casks, being first well roused with the remaining part of the yeast, in the manner mentioned in the preceding section.

By the description given of the origin of yeast, it appears that it is formed rather of the coarser oils of the worts. If the cleansing is not done when the head is sunk down to half the greatest height it rose to, by falling lower, some part of these coarser oils return into the beer, then under fermentation, and gives it a flat, greasy taste, technically termed *yeast bitten*. When, on the contrary, beers or ales are removed too soon from the first tumultuous fermentation, for want of having been sufficiently attenuated, and from not having deposited their lees, nor thrown up in flowers their coarser oils, they are less vinous, than otherwise they would have been, appear heavy, aley, and are said *not to have their body sufficiently opened*.

The fermentation of common small beer is, through necessity, carried on so hastily, that it is hardly possible to wait for the signs, which direct the cleansing of other beers. This drink being generally brewed and fermented within twenty four hours, its state, with regard to fermentation, is best judged of, by the quantity of its froth or head at the time of cleansing, which, in proportion to the heat of the air, may be determined by the following table.

A TABLE *shewing the depth of head, which common small beer should have to be properly cleansed, in every season of the year.*

Heat of the air.	Head on the beer in the tun.
25 Degrees	6 inches.
30	5
35	$4\frac{1}{4}$
40	$3\frac{1}{2}$
45	$2\frac{3}{4}$
50	2
55	$1\frac{1}{2}$

60	1
65	¾
70	½
75	¼
80	just taken.

As it is chiefly by the action of the air that wines are formed, if we contrive to shift this powerful agent on the surface of a must under fermentation, and to convey it more forcibly and hastily into the wort, its efficacy will be renewed, the fermentation accelerated, the liquor quickly become transparent, and soon be brought to the state of maturity age might slowly make it arrive at.

Amber, or pale ales, require the hottest extracts pellucidity admits of to be made strong, and at the same time soft and smooth to the palate; but, as ales do not admit of any large quantity of hops, which would alter their nature, there is a necessity to perform hastily the act of fermentation, and to carry it on to a higher degree than is common in other malt drinks. The method of exciting and conducting repeated fermentations, with success, is perhaps not only the most difficult, but the most curious, part of the process, I shall therefore conclude, with an account of it, what I have to say with regard to the practice of fermentation.

The amber wort being let down, at its proper degree of heat, into the fermenting tun, out of the whole quantity of yeast allowed for this drink, in the table, page 317, one seventh part must be kept to be used as hereafter shall be mentioned. Suppose the heat of the air is at 40 degrees, and eight quarters of malt have been brewed for this purpose; the whole of the yeast required is seven gallons, from which one is to be reserved.[39] Of the six remaining gallons, one half, or three, are to be put to the wort oh its first coming down, when the whole must be well roused, or mixed, thoroughly to disperse the enlivening principle the yeast conveys, hereby to prevent putrefaction, or foxing in any part, and of the last three gallons, about three quarts must be added to the drink, every twelve hours, until it ferments to the highest pitch of the period mentioned in article 5, page 319. This successive putting in of yeast is called[40] *feeding the drink*; before and about the time the head is got to this height, all the dirt or foul yeast, that rises on the surface, must be carefully skimmed off; it is easily distinguished from the pure white froth, by its color, and by the sinking of the head occasioned by its weight. Length of time might attenuate some of these coarser oils, in a less artificial fermentation, but as this help is not to be waited for, and every obstacle to pellucidity must be removed, the brewer's attention to this point cannot be too great.—The head of the drink having reached its utmost height, the reserved gallon of yeast is to be used, in order to give to the ale a sufficient power to bear the repeated fermentations it is to undergo, by being beat in, every two hours, with a jett or scoope, for one quarter of an hour, so that the head on the drink is each time to be reduced to the least height it is capable of. This striking in being continued, the drink will periodically require it, and be damaged if it be neglected. After it has undergone more or less of these fermentations, in proportion to the heats of the worts and of the air, the brewer is carefully to observe, when the head ceases to rise to its accustomed height, and then to examine the drink, by having the jett filled with it at the bottom, and brought through the whole body to the top, a small part of which being poured in a handgatherer, he will see whether the lees form themselves in large white flakes, and readily subside, and be informed, by the taste, whether the sweet of the wort is gone off, and the ale become vinous. If these two circumstances concur, the drink is to be beat in with the jett as before, but not roused as porter or other beers are; for the lees, which in this drink are in greater quantity, would, by this management, so intimately be mixed with it, as with difficulty to separate themselves again, if at all. It is then time to cleanse it; but the casks, at all times, more especially in summer, must be well filled up with clean drink, that is, part of the very drink, which was cleansed, avoiding that produced in the stillings, as this, for want of standing a sufficient time, is always yeasty, and the yeast, being greatly attenuated by the working of the drink, easily dissolves in the ale, and renders it foul and ill-tasted.

As the right forming of amber ales is looked upon to be the highest pitch of the art of pale beer brewing, I have dwelt longer on this article than otherwise it might seem necessary, to shew the connexion there is between every sort of malt liquors: but it should be observed, the same method of fermenting it, is to be practised both winter and summer, varying only the quantities of yeast in proportion to the season; for where, in winter time, this drink is fed with three quarts of yeast every twelve hours, half a gallon will answer the same intent in summer. Upon the whole, the process is contrived to accelerate fermentation, yet, the more coolly and gently it is performed, the better will the ales be. I have before hinted, if Madeira wines were fermented in this manner, they would sooner become fit for use, more especially as they need no ferment to excite them. However, this method of forming drink to be soon fit for use, has, either through interest or prejudice, been taxed with being unwholesome, but upon what grounds, I must confess I could never yet discover, as no reason of any moment has ever been alledged for this assertion.

SECTION XVII.

Of the signs generally directing the processes of Brewing, and their comparison with the forgoing Theory and Practice.

WE have now brought our barley wines into the casks, and this on principles, it is thought, agreeable and consonant to each other. As the charge of novelty may be alledged, to invalidate what has been offered, it is but just to pay so much regard to a long, and, upon the whole, successful practice, as to recite, if not all, at least the principal maxims and signs in brewing, which hitherto have guided the artist. By comparing these with the present method, they will not only illustrate each other, but perhaps cause both to be better understood; and though, with respect to the art itself, this may be thought rather a curious than an instructive part, yet we may learn, from hence, that such practice, which long experience has proved to be right, will always correspond with true theory.

1. *When a white flour settles, either in the underback or copperback, which sometimes is the case of a first extract, it is a sure sign such an extract has not been made sufficiently hot, or, in technical terms, that* the liquor has been taken too slack.

Malt, when dried, has its oils made tenacious, in proportion to the power of heat it has been affected with; the grain, though ground, if the water for the extraction is not at least as hot as what occasioned this tenaciousness, must remain in great measure undissolved in the first extract, and deposit itself as just now was mentioned.

2. *The first extract should always have some froth or head in the underback.*

The oils and salts of the malt, being duly mixed, form a saponaceous body, the character of which is that, on being shook, it bears a froth on its surface.

3. *The head or froth in the underback appearing red, blue, purple, or fiery, shews the liquors to have been taken too hot.*

The hotter the water is, when applied to the malt, the more must the extract abound with oils, and consequently be more capable to reflect colors in a strong manner. But how precarious this method of estimating the quality of an extract is, in comparison to that which the thermometer affords, will appear from the following observation of Sir Isaac Newton: "Saponaceous bubbles will, for a while, appear tinged with a variety of colors, which are agitated by the external air, and those bubbles continue until such time as, growing excessive thin, by the water trickling down their sides, and being no longer able to retain the enclosed air, they burst." Now as these bubbles vary in their density, in proportion to their duration, the colors they reflect must continually change, and therefore it is not possible to form an accurate judgment of the condition and saponaceousness of the extracts, by the appearance of their froth.

4. *When the grist feels slippery, it generally is a sign that the liquors have been taken too high.*

This appearance proceeds from an over quantity of oil being extracted, and is the effect of too much heat.

5. *Beer ought always to work kind, out of the cask, when cleansed, but the froth, in summer time, will be somewhat more open than in winter.*

The higher and hotter the extracting water is, the more oils doth it force into the must; when a wort is full charged with oils, the fermentation is neither so strong nor so speedy, and consequently the froth, especially the first, is thin, open, and weak. This improves as the liquor is more attenuated, and heat, which expands all bodies, must rarify the yeasty vesicles, the principal part of which is elastic air; but this open head, even in summer time, improves to one more kind, as the first, the most active period of fermentation, draws nearer to its conclusion.

However vague and indeterminate these signs are, it would not be impossible to bring them to some degree of precision; but, upon the whole, this method would increase our difficulties, and yet, as to certitude, be inferior to the rules we have endeavoured to establish, we think it unnecessary to pursue any farther a research most likely neither entertaining nor useful.

SECTION XVIII.

An enquiry into what may be, at all times, a proper stock of Beer, and the management of it in the cellars.

THE business of a brewer is not confined to the mere manufacture of his commodity; his concerns, as a trader, deserve no less regard, and, in a treatise like this, should not be entirely omitted.

As it is a fault not to have a sufficient stock of beers it the cellars, to serve the customers, it is one also to have more than is needful. By the first of these errors, the beers would be generally new and ill disposed for precipitation; by the other, quantities of stale beer must remain, which, becoming hard, will at last turn stale, and be unfit for use, unless blended with new brewed beers, to their detriment. These faults, if continued, may in time affect a whole trade, and ought therefore carefully to be avoided. For these reasons, the whole quantity to be moved, or expected to be supplied from the brewer's store cellars, during the space of one twelvemonth, should be calculated, as near as possible; half this quantity ought to be the stock kept up from November to May inclusive, and nearly one third part thereof be remaining in September. From hence a table may be formed, by which it will be easy, at one view, to know the quantity that should be maintained at every season of the year, and to avoid almost every inconveniency, which otherwise must arise. Suppose, for example, the number of casks expected to be moved in a year, to be 320 butts, and 248 puncheons, the store cellars ought to be supplied, as to time and quantity, in the following proportion.

	Butts.	Puncheons.
January	160	124
February	160	124
March	160	124
April	160	124
May	160	124
June	146	113
July	133	103
August	120	93
September	107	82
October	133	103

November	160	124
December	160	124

After beers have been started in the cellars, the casks should be well and carefully stopped down, as soon as the repelling force of fermentation is so much lessened, as not to be able to oppose this design. Otherwise the elastic air, which is the vivifying principle of the drink, being lost, it would become vapid, and flat; and if left a long time in this condition, perhaps grow sour.

It has already been observed, that cellars, in winter, are more hot than the exterior air by 10 degrees, and more cold in summer by 5 degrees. But besides this general difference, repositories of beer vary surprisingly in their temperature; from the nature of the soil in which they are built, from their exposition to the sun, or from other incidental causes. As heat is a very powerful agent in accelerating fermentation, it is by no means surprising, not only that some cellars do ripen drinks much sooner than others, but also that a difference is often perceived in the same cellar. The persons entrusted with the choice of beers, with which the customers are to be served, should not be satisfied to send out their guiles in the progressive order in which they were brewed, but ought, on every occasion, to note any alteration that happens in the drink, as this is doing justice both to the commodity, and to the consumer, who has a constant right to expect his beer in due order.

SECTION XIX.

Of Precipitation, and other remedies, applicable to the diseases incident to Beers.

NO accident can be so detrimental as leaky or stinking casks, which lose or spoil the whole or part of the contained drink. The necessity of having, on these occasions, a remedy at hand, was undoubtedly the reason, why coopers were first introduced in store cellars. Constant practice might have qualified their palates so as to make them competent judges of the tastes of wines and beers, and to enable them to know which were the fittest for immediate use. The preparing or forcing them for this service, was a matter, which the profit gained thereby made them ready enough to undertake. Chymists, whom they consulted on this occasion, gave them some informations, from whence the coopers became the possessors of a few nostrums, the effects of which they were supposed to have experienced. But, ignorant of the causes of most, if not all the defects they undertake to remedy, and unacquainted with the constituent parts of beers, it is not to be expected that their success should be constant and uniform. The brewer, earnest to do his duty, and to excel, ought to keep a particular account of every brewing; by this means he best can tell how he formed the drink, and ought consequently, in any disorder, to be prepared to direct the properest remedy.

The intent of this treatise has been to discover the means by which errors may be avoided. Chymical applications are intended to remedy those errors, which may be occasioned either by carelessness or accident. The wholesomeness or propriety of the applications, which will be indicated, must be left to the judgment of my readers; it is most likely that there is sufficient room for improvement, and we might expect it from those, whose profession it is to study every thing, that may be conducive to the safety of mankind.

Whatever vegetables wines are produced from, whenever they deviate from the respective perfection, a well-conducted fermentation might have made them arrive at, they may be said to be distempered. Foulness, or want of transparency, is not the least evil, but, according to its degree, it obtains various appellations, and requires different helps. From what has been said, nothing can be more plain, than that it is always in our power to form beers and ales, which will be bright. Yet porter or brown beer is constantly so brewed as to need precipitation: the reasons for this management have before been offered. Were we to wait till the liquor became transparent by age, a more real disorder would ensue, that of acidity. Precipitation is then serviceable, especially when beers are to be removed from one cellar to another, a short space of time before they are to be used. By being shook, and the lees mixed with the liquor, a strong acid taste is conveyed therein, and the power of subsiding, which is wanted, renders the forcing them, in that case, of absolute necessity. In beers brewed with liquors sufficiently heated, no flatness is occasioned thereby; as the case is, under like circumstances, with liquors produced by low extracts, from grain not sufficiently dried. The degree of foulness in porter should however be limited; its bounds ought not to exceed the power of one gallon of dissolved isinglass, to a butt. Isinglass is dissolved in stale beer, and strained through a sieve, so as to be of the consistence of a jelly. The beer is set in motion with a stick, which reaches one third part down the cask, before and after this jelly is put in; and a few hours should be sufficient to obtain the desired effect. We have before observed, that this quantity of jelly of isinglass is equal to a medium of 10 degrees dryness in the malt, and heat of the extracts. When the opacity exceeds this, the liquor is termed *stubborn*; the same quantity of dissolved isinglass repeated, is often sufficient, if not, six ounces of the oil of vitriol are mixed with it. An effervescence is, by this addition, produced; the oils of the drink become more attenuated, and the weight added to the precipitating matter, is a means to render it more efficacious. Instead of the oil of vitriol, six or eight ounces of the concrete of vitriol, pounded and mixed with the isinglass, are sometimes used with success.

A foulness in beer beyond that which is called *stubborn*, gives to the drink the denomination of *grey beer*. This arises from the oils which float upon the surface, and which the liquor has not been able to absorb. In this case, the same methods as before mentioned are repeated; the quantity of dissolved isinglass is often increased to three gallons, that of vitriol to more than 12 ounces, and sometimes a small quantity of *aqua fortis* is added to these ingredients.

The next stage of opacity is *cloudiness*; when the cooper confesses that the distemper exceeds the power of his menstruums, and that his attempts extend no farther than to hide the evil, tournsol and cochineal, were they not so expensive, might in this case be used with success; but what is less known, and would greatly answer the intent of hiding the dusky colour of the drink, is madder;—about three or four ounces of this is the proper quantity for a butt of beer. Calcined treacle, by the coopers called blacking, from its acidity, is of some small service, for, by coloring the drink, it somewhat lessens the grey hue thereon; a quart is generally used in a butt; and, to prevent the defect in the beer being noticed by the consumer, the practice is to put thereon what is called *a good cauliflowered head*. This might be done by using as much pounded salt of steel as will lay upon a shilling; but the difference in price between this salt and copperas makes the last generally to be preferred. The strong froth on the top of the pot, and that which foams about it, together with somewhat of a yellow cast, are often mistaken for the signs of a superior merit and strength, though, in fact, they are those of deceit. A little reflection that the natural froth of beer cannot be yellow, nor continue a long time, especially if the liquor has some age, would soon cure mankind of this prejudice. Cloudy beers, under these circumstances, though not cured, are generally consumed.

Beers become *sick*, from their having so large a portion of oils, as to prevent the free admission of the external air into them. The want of this enlivening element makes them appear flat, though not vapid. Such beers should not, if possible, be brought immediately into use, as age alone would effect their cure. But when this cannot be complied with, every means that will put the beer upon the fret, or under a new fermentation, must be of service. By pitching a butt head over head, the lees of the beer, which contain a large proportion of air, being mixed again with the drink, help to bring on this action, and to remove the *sickness*.

Burnt hartshorn shavings, to the quantity of two-penny-worth, put into a butt, are often of use.

Balls made with eight ounces of the finest flower, and kneaded with treacle, convey likewise air to the drink, and promote its briskness.

Beers, by long standing, often acquire so powerful an acid, as to become disagreeable. The means of correcting this defect is by alkaline, or testaceous substances, and in general by all those which have the property of absorbing acids. To a butt of beer in this condition, from four to eight ounces of calcined powder of oyster-shells may be put, or from six to eight ounces of salt of wormwood. Sometimes a penny-worth or two of whiting is used, and often twenty or thirty stones of unslacked lime; these are better put in separately, than mixed with the isinglass.

From two to six pounds of treacle used to one butt of beer, has a very powerful effect, not only to give a sweet fulness in the mouth, but to remove the acidity of the drink. Treacle is the refused sweet of the sugar baker, part of the large quantities of lime used in refining sugars, undoubtedly enter in its composition, and is the occasion of its softening beers.

In proportion as beers are more or less forward, from two to four ounces of salt of wormwood and salt of tartar, together with one ounce of pounded ginger, are successfully employed. All these substances absorb acids, but they leave a flatness in the liquor, which in some measure is removed by the use of ginger.

Sometimes, in summer, when beer is wanted for use, we find it on the fret; as it is then in a repelling state, it does not give way to the finings, so as to precipitate. For this, about two ounces of cream of tartar are mixed with the isinglass, and if not sufficient, four ounces of oil of vitriol are added to the finings next used, in order to quiet the drink.

Some coopers attempt to extend their art so far as to add strength to malt liquors; but let it be remembered, that the principal constituent parts of beer should be malt and hops. When strength is given to the liquor by any other means, its nature is altered, and then it is not beer we drink. Treacle in large quantities, the berries of the *Cocculus Indicus*, the grains of paradise, or the Indian ginger pounded fine, and mixed with a precipitating substance, are said to produce this extraordinary strength. It would be well if the attempts made to render beers strong by other means than by hops and malt, were to be imputed to none but coopers; Cocculus Indicus, and such like ingredients, have been known to be boiled in worts, by brewers who were more ambitious to excel the rest of the trade, than to do justice to the consumers. Were it not that pointing out vice is often the means to forward the practice of it, I could add to this infamous catalogue, more ingredients, it were to be wished practitioners never knew either the name or nature of, for fining, softening, and strengthening.

Formerly brown beers were required to be of a very dark brown, inclinable to black. As this color could not be procured by malt properly dried, the juice of elder berries was frequently mixed with the isinglass. This juice afterwards gave way to calcined sugar; both are needless, as time and knowledge remove our prejudices, when the malt and hops have been properly chosen; and applied to their intended purpose.

Such are the remedies chiefly made use of for brown beers. Drinks formed from pale malts are always supposed to become spontaneously fine, and when they are so, by being bottled, they are saved from any farther hazard. As it is impossible for any fermented liquor to be absolutely at rest, the reason of beers being preserved by this method, is, thereby they are deprived of a communication with the air, and, without risk, gain all the advantages which age, by slow degrees, procures, and which art can never imitate. Were we as curious in our ales and beers as we are in the liquors we import, did we give to the produce of our own country the same care and attendance which we bestow on foreign wines, we might enjoy them in a perfection at present scarcely known, and perhaps cause foreigners to give to our beers a preference to their own growth.

SECTION XX.

OF TASTE[41].

DOCTOR GREW, who has treated of this matter, divides taste into simple and compound; he mentions the different species of the first, and calculates the various combinations of the latter, the number of which exceeds what at first might be expected. Without entering into this detail, I think that the different tastes residing in the barleys, or formed by their being malted, and brewed with hops, may be reduced to the following; the acid, which is a simple taste; the sweet, which is an acid smoothed with oils; the aromatic, which is the compound of a spiritous acid, and a volatile sulphur; the bitter, which, according to our author, is produced by an oil well impregnated either with an alkaline or an acid salt, shackled with earth; the austere, which is both astringent and bitter; and, lastly, the nauseous and rank, which is, at least in part, sometimes found in beers, which have either been greatly affected by fire, or, by long age, have lost their volatile sulphurs; and have nothing left but the thicker and coarser oils, resembling the empyreumatic dregs of distilled liquors not carefully drawn.

The number of circumstances on which the taste of fermented liquors depends, are so various, that perhaps there never were any two brewings, or any two vintages, which produced drinks exactly similar. But as, in this case, as well as in many others, the varieties may be reduced under some general classes; the better to distinguish them, let us enquire which taste belongs to different malt liquors, according to the several circumstances in which they are brewed.

In beers and ales, the acid prevails in proportion as the malt has been less dried, and heat was wanting in the extracting water. The sweet will be the effect of a balance preserved between the acids and the oils. When, by the means of hotter waters, oils more tenacious are extracted from the grain, whereby the more volatile sulphur is retained, the taste becomes higher in relish, or aromatic. If the heat is still increased, the acids, and the most volatile oils, will in part be dissipated, and in part be so enveloped with stronger oils, as the bitter of the hops appears more distinct. A greater degree of fire will impress the liquor with an austere, rough, or harsh taste; and a heat beyond this so affects the oils of the grain, as to cause the extracts to be nauseous to the palate. Besides these, there may be other causes which produce some variation in taste; as a superior dryness in the hops; an irregularity in the ordering of the heat of the extracts; too great an impetuosity or slowness in the fermentation; the difference of seasons in which the drink is kept; but as these causes affect the liquor, in a low degree, in comparison to the drying and extracting heats of the grain, an enquiry into their consequences is not absolutely material.

Beers or ales, formed of pale malt, in which a greater portion of acids is contained, with less tenacious oils, are not only more proper to allay thirst, but in general more aromatic than brown drinks. The oils of these last, being, by the effect of fire, rendered more compact, and more tenacious of the terrestrial parts raised with them, are attended with something of an austere and rank taste. This seems to be the reason why brown beers require more time, after they have been fermented, to come to their perfection. The air, by degrees, softens and attenuates their oils, and, by causing the heterogeneous particles to subside, makes them at last, unless charring heats have been used, pleasing to the palate, whereas they were before austere, rank, and perhaps nauseous.

By means of the thermometer, we have endeavoured to fix the different colors of malt, the duration of the principal sorts of drink, and the tendency each has to become transparent. The same instrument cannot probably have the same use, when applied to distinguish the different tastes, as these depend on a variety of causes not easy to be ascertained. Yet something of this nature may be attempted, upon the following principles.

As the chief circumstance which produces a variety of tastes in malt liquors, is fire or heat acting on the malt and hops, and the effect of the air, put in motion by the same element, the table here subjoined may point out what tastes are in general occasioned by the combination of these two causes.

A TABLE *determining the tastes of Malt Liquors.*

Heat of the air.	Dryness and extracting heat.	Predominant tastes.
80°	119°	Acid.
76	124	Ac. ac. sweet.
73	129	Ac. sw.
70	134	Ac. sw. sw. bitter.
66	138	Sw. sw. bitter.
63	143	Sw. bit.
60	148	Bit. bit. aromatic.
56	152	Bit. arom.
53	157	Bit. arom. austere.

50	162	Arom. aust. aust.
46	167	Aust. aust. nauseous.
43	171	Aust. nau.
40	176	Nauseous.

The first column of the table shews the fermentable degrees reversed, as the hotter the season is, the more fermented drinks tend to acidity, the direct contrary of which is the consequence of an increase in the heat, malt or hops are dried or extracted with.

The assistance of this table, though small, ought perhaps not to be entirely slighted, as it seems at least to shew that the useful is seldom separated from the elegant, and that a medium between extremes is most agreeable both to the operations of nature, and the constitution of our organs.

The impressions of tastes are less in proportion as the drinks are weak. The strongest wine yields the most acid vinegar. Time wears away this acidity much sooner, than it doth the nauseousness occasioned by vehement heats. This circumstance shews how necessary it is, in the beginning of the process of brewing, to avoid extracts which are too weak, as from hence, in its conclusion, such would be required whose great heat would render the drink rank and disagreeable. That proportion between the salts and the oils, which constitutes soundness and pellucidity, is most pleasing to the taste, and seems to be the utmost perfection of the art. As the sun never occasions a heat capable of charring the fruits of the vine, we never meet with wines endued with a taste resembling the empyreumatic, which we have here represented. This error, being inexcusable in any liquor, ought carefully to be guarded against, and, from what has here been said, we should learn this important truth, that nature is the best guide, and that, by imitating, as near as possible, her operations, we shall never be disappointed in our ends.

APPENDIX.

THOUGH this work has already been carried to a great length, I hope those of my readers, who may have done me the honor to go attentively through the whole of it, will pardon me the addition of a few incidental thoughts and queries. The chain of arts is so well connected, that researches originally intended for the illustration of any one of them, can hardly fail of throwing some light upon others.

1. The seed of plants cannot be put in a fitter place, for perfect vegetation, than when buried under ground, at a depth sufficient to defend the young shoots from the vicissitudes of heat and cold, and the disadvantage of too much moisture. The manuring of the earth, and the steeping the seed into solutions of salts, have been found, in some cases, to increase the strength of the grain, to correct its original defects, and to prevent the noxious impressions of a vicious ground. Plants are made to germinate in water alone, and this experiment so successfully carried on every winter, in warm apartments, may still be improved by dissolving salts in the water.—Could the barley used for malting be put in the ground, its growth would be more natural, and its oils becoming more miscible with water, by the saline nourishment derived from the earth, might yield more vinous, more strong, and more lasting liquors. But as this method is impracticable, would it be impossible to increase the efficacy of that which is used? Consult Home on agriculture: might not either nitre or salt petre be added to the water, with which the grain is moistened? are they not used with success to manure land? Are not solutions of them in water employed by the farmer to steep his sowing seed? I barely mention these as some of the substances, that might be employed in the malting of barley, and am far from thinking there are none other. Perhaps different salts should be used, according to the nature of the soil, from which the corn was produced; but a variety of experiments seems to be required, in order to discover how far art might in this case imitate and improve nature.

2. A small quantity of malt, at all times, but especially when brewed in large vessels, parts too readily with the heat which extraction requires; and, on the contrary, if the quantity of malt be very great, the heat may not be uniformly spread. A forward beer inclinable to acidity is often the result of too short a grist; a thick, stubborn, and rank liquor many times is produced from too large a one. Every advantage may be had in brewing, properly, five or six quarters of malt; it is difficult to succeed if the number exceeds fifty.

3. The strong pungent volatile spirit, which exhales from a must, when under full fermentation, has been supposed to be a loss, which might be prevented; and accordingly attempts have been made to retain these flying impetuous

particles, by stopping the communication between the atmosphere and the fermenting drink. That there is a dispersion of spirits is beyond doubt, and that these exhaling vapors consist of the finest oils, which the heat forces out of the must, is equally certain. But this loss seems to be abundantly repaid by the stronger oils, which the same degree of heat attenuates and substitutes, in a larger quantity, to the former. The last oils could never come under the form of a vinous liquor, but by a power, which sooner or later dissipates some of the first. Pale ales or amber not only lay, for many days, exposed to the open air, but suffer, by the periodical renewal of the action of the air, every two or four hours, a much more considerable loss of spirits, than when fermentation is carried on uniformly. Yet experience shews, that so many oils are, by this method, attenuated, that the strength acquired greatly surpasses that which is lost.

4. The practice of fermenting *by compression*, recommended to distillers, seems, on this account, less useful, than might be concluded from theory alone; the intent of the distiller, as well as of the brewer, is to extract the greatest quantity of spiritous oils. It is impossible to ferment a must *in vacuo*; air is absolutely necessary to carry on this operation, even a superabundant quantity of oils admitted into the must, by obstructing the free admission of the air, impedes fermentation, prevents the wine from reaching pellucidity, and sometimes is the occasion of its becoming putrid.

5. When the purest spirit is intended to be drawn from the grain, the fermented wash ought to be suffered to settle, till it becomes transparent. The dispatch, with which the distillery is generally carried on, often prevents this useful circumstance taking place, and occasions a want of vinosity in the liquor. In many cases, the extraordinary charges of extracting the grist from malted corn, in the manner, which has been directed for drinks intended a short space to be kept, and of suffering the fermented wash to be meliorated by time, until it becomes vinous and spontaneously transparent, might be abundantly repaid. Yet, if hurry must be a part of the distiller's business, he should at least make such extractions as admit of the speediest fermentation and the readiest pellucidity. He cannot expect corn spirits to equal the brandies of France, unless his worts are similar to the wines distilled in that kingdom, where those used for this purpose are weak, fine, and tending to acidity.[42] He would therefore secure to himself the greatest probability of success, if he employed only malted corn in his grist, this of the best kind, well germinated to form a saccharine basis, slack dried, and resolved, with weak extracts, to preserve into the must a proper proportion of vinosity. If he intended this wash to be formed into a pure spirit, it should be allowed time to become transparent; he might regulate his extracts by such heats as have been fixed for common small beer, brewed when the heat of the air is at the lowest fermentable degree, though perhaps heats less than these, when

dispatch is required, might better answer his purpose, especially as the length used in the distillery is nearly the same with that which brewers use for the liquor here referred to. With hot waters to attempt to force from the grain more strength or more oils, than such as will form a clean tasteless spirit, is, in the distillery, a real loss and a fundamental error. By too strong heats, more oils are forced into the must than can be converted in spirits; and fermentation being, by this over charge, in some measure, clogged and impeded, a less yield is made, and the liquor obtained of a rank and often empyreumatic taste.

6. Why are the brandies of Spain inferior to those prepared in France? The wines of the last country are the growth of a weaker sun; they contain no more oils than can be assimilated by fermentation, and form a clean, dry, nutty spirit. The Spanish wines abounding with more oleaginous than acid parts, this over proportion becomes not only useless, but hurtful in the still, and produces the rankness observed in Spanish brandies. The cleanness of the spirit arises, in great measure, from the weakness of the must, and its vinosity from a less proportion of oils to the salts. This seems to be the reason why the most grateful spirits are produced from wines unable to bear the sea, or to be long kept.

7. The native spirits of vegetables, says Boerhaave, are separated by heats between 94 degrees, and 212. To obtain the whole of these, the fire must be gradually increased; for a superior heat dissipates the spirits raised by an inferior one. Such parts as might be obtained by 100 degrees, are lost if the heat applied be much greater. It is true, the parts of vegetables immersed in water, cannot so easily be dissipated as if they were in open air, yet, by the rarefaction of the liquid, a proportional evaporation, however small, must ensue, or the oils raised by a greater heat may so effectually envelope the finer ones, as to make them hardly perceptible either to our smell or taste. Thus, though heated water is able to extract all the virtues residing in the vegetables, the different application of the fire will alter not only their proportions, but their properties also, when we consider that pure spirit of wine boils at so low a heat as 175 degrees. If the above principles be true, that surely must be the cleanest spirit which is brought over in the slowest and coolest manner; and it is more than probable, if the rules here laid down be put in practice, the grain of England will be found to yield spirits that may vie with the brandies of France, be more pure than those of the Indies, and excel those of Holland.

8. The vinegar maker is equally concerned with the distiller in the brewing process. Vinegar is produced in the last stage of fermentation, when a gross, tartareous, unctous matter, consisting of the coarser oils extracted either from the grain or the grapes, generally falls to the bottom of the liquor, and no longer prevents its acidity, or affects its flavor. Though the best vinegar

proceeds either from the strongest wines or beers, this strength consists in the quantity of fermentable principles, and not in that of mere oleaginous parts. By properly adapting the extracting waters, this hurtful impediment may be removed, and the vinegar from malt liquors become as neat and as strong as that which is made from wine.

9. As the acid taste of vinegar is the effect of a continued fermentation, many people have thought it immaterial how speedily the first parts of the operation were carried on. But violent fermentations not only dissipate some of the fine oils, which should be retained in the vinegar, but also cause the must to tend towards putrefaction. Boerhaave, after he has directed a frequent transvasion of the liquor, observes that, whenever the weather or the workhouse is very hot, it is often necessary to fill the half emptied vessels every twelve hours, not only to procure a supply of acids from the air, but also to cool the wine, and check the too violent fermentation, which arising in the half full casks, might dissipate the volatile spirits, before they are properly secured and entangled by the acid. Hence the liquor might be sour indeed, but at the same time flat, and would never become a sharp and strong vinegar.

10. Application and uses have frequently been found for materials, which before were supposed to be of no value. The grains, after the brewer has drawn his worts out of them, are generally used for the feeding of cattle; but I do not know that hops, after boiling, have been employed to any purpose. Is there nothing more left in this vegetable, after it has imparted the virtue wanted to the beer? All plants burnt in open air yield alkaline salts, though in a greater or less quantity, according to the quality of the plants. Boerhaave says that those which are austere, acid, or aromatic, yield in their ashes a great abundance of salts, and these being put in fusion, and mixed with flint or sand, run into glass. Hops thrown, after decoction, in no great quantity on the fire, cause the coals to vitrify, or as it is generally termed, to *run into clinkers*. If therefore the remains of the hops were burnt in open air, or in a proper furnace, it seems most likely that no inconsiderable357 quantity of somewhat like pot ashes might be obtained, and this, considering the many tun weight of hops employed in large cities, and thrown away as useless, might become an object of private emolument to the brewer, and of public benefit to the kingdom.

FINIS.

FOOTNOTES:

<u>1</u> Vide Dr. Pringle's experiments in his book of observations on the diseases of the army, p. 350, 351 & seq.

<u>2</u> There is a very singular exception in regard to iron itself, in this respect. It is only a certain degree of heat that expands this metal; (and that much less than any other either more or less dense) when melted, it occupies a less space than when in a solid form. This ought to caution us against an entire dependence on general rules, by which nature doth not appear to be wholly restricted. See Mem. de l'Acad. des Scienc. p. 273.

<u>3</u> See Dr. Lewis's Philosophical Commerce of Arts, p. 42.

<u>4</u> See Martine's Dissertation on Heat. What the degree of cold was which fixed mercury at St. Petersburg, I do not recollect.

<u>5</u> It requires seven or eight days. (See Dissertation sur la glace par Mons. de Mayran.) Paris edition, 1749. Page 191.

<u>6</u> Lately, indeed, by such intense cold as can only be procured with the greatest art, and in the coldest climates, mercury is said to have been stagnated, or fixed.

<u>7</u> By Dr. Hales's experiments made for discovering the proportion of air generated from different bodies, it appears that raisin wine, absorbed, in fermenting, a quantity of air equal to nearly one third of its volume; and ale, under the like circumstances, absorbed one fifth.

<u>8</u> In the northern part of England, the usual time of steeping barley in the cistern is about 80 hours.

40 bushels of barley wetted 1 hour, will guage then in the couch 40 bushels, that is, if drained from its exterior moisture.

40 bushels	20 hours,		42½ bushels.
40 bushels	40 hours,		45 bushels.
40 bushels	60 hours,		47½ bushels.
40 bushels	80 hours,		50 bushels.

Here the barley is supposed to be fully saturated with the water; and these 40 bushels of barley, guaged (after 80 hours wetting in the cistern) in the couch, will be 50 bushels; but when again guaged on the floor, from the effect of the roots, and sometimes the shoots, occasioning the corn to lie hollow, here the 40 bushels of barley will shew as 80 bushels. Vide Ramsbottom, page 113, &c.

9 Boerhaave Elem. of Chym. Vol. I. p. 195-199. Exp. 8, 9, 10, 11, 12, and 13.

10 When the medium heat of the dryness of the malt, and of the heat of the extracts, are so high as to require the liquors to be forced or precipitated, in order to become pellucid, part of the oils which supported them sound, being carried down by the precipitant, they will be less capable of preserving themselves, after having been precipitated, than they were before.

11 I chose this manner of expressing the quantity of moisture received in ground malt from the air, as it is the most easy for the direction of the first extract.

12 Part I. Sect XII. p. 124.

13 See page 56.

14 For the properties answerable to the degrees, see page 124.

15 It may be observed that, in the first and last degrees for drying malt, sometimes we say one degree more, sometimes a degree less.—The experiments we have made do not admit of a geometrical exactness, nor does the practice of brewing require it; small errors in beers are effectually removed by age, and these variations have often been adopted in the tables, for the conveniency of dividing into whole numbers.

16 See p. 124.

17 *Purl*, is pale ale, in which bitter aromatics, such as wormwood, orange peel, &c. are infused, used by the labouring people, chiefly in cold mornings, and a much better and wholesomer relief to them, than spiritous liquors.

18 152, to which 2 degrees must be added, for what is lost in the extracts coming away, or 154 degrees, being the heat of the mash for keeping small beer, after amber; as this number is less than 166 degrees, the last mash of the amber, consequently, in the computation made, to find how much of the quantity of the liquor used, is to be made to boil, to give the true degree of heat to the mash of small, the difference of heat required in this mash, 154, and the heat of the goods 162 or 8, is to be multiplied by the volume of the goods, and the product in this case subtracted; whereas, in the operations for brewing, whose heat gradually increased every mash, it is to be added.

19 We had rather attribute to this cause, the inferior quality of the Worcestershire hops, than to what is reported. That some planters in that county suffer their hops to be so ripe on the poles, that they become very brown before they are gathered: to recover their color, on the fire of the kiln

they strew brimstone, which brings them to a fine yellow; the dryness and harshness this acid occasions, they correct by sprinkling the hops with milk, from whence they bag closer, and require little straining, but two ingredients more pernicious to the forming good beers, perhaps, could not have been thought of, than milk and brimstone.

20 This rule only takes place for such climates as are of the same heat with ours; for when drinks are brewed to be expended in more southern countries, or to undergo long voyages, twenty pounds of hops to one quarter of malt have been used with success.

21 If, of the whole quantity of hops grown in one year, one half is put into bags, whose tare is one tenth of their whole weight, and the other half is put in pockets, whose tare is one fortieth of their whole weight; if the excise office allows one tenth for tare upon the whole, and the excise or weighing officers, are content with one ninth, as by their marks, and the weight when sold to the brewer, appears to be the fact; then somewhat like one twentieth part more hops are grown, than what pays duty, or than the excise officers report to be the case.

22 Forty shillings per hundred weight, are supposed to be the mean difference between new and old hops, and ought to be estimated in proportion to the quantity of old left in hand, and that of new hops grown, in order to ascertain the value of the last.

23 B. stands for Barrels, F. for Firkins, G. for Gallons.

24 When there are but two worts in brown strong, keeping strong, keeping pale small, or common small, the boiling is to be observed as marked for the second and third worts.

25 The small cask, called a *pin*, is one eighth part of a barrel.

26 By new malt, I understand such, as has not lost the whole of the heat received on the kiln, and by old, such as is of equal heat with the air, or such which has laid a sufficient time to imbibe part of its moisture.

27 At the time when the first edition of this work was published, porter or brown beers were brewed with very high dried malts; experience has shewn to the generality of the trade and to the author, this practice to be erroneous, the reasons why have before, and perhaps hereafter will again, be spoken of. In compliance with this improvement (though between the two proposed brewings, so great a variety will not appear) I have founded my calculations for porter, on malts dried so as best will answer this purpose.

28 B. stands for barrels, F. for firkins, G. for gallons, and the numbers past the comma, where the inches are expressed, for decimals; 34 gallons are here

allowed to the barrel, in compliance to the excise gauging, as these calculations were made without the bills.

29 The half degree omitted in this mash will be added to the next.

30 Different quantities of water are differently affected by the same portion of fire; when the ebullition is just over, and the surface of the liquor is become smooth; if some of it is, by a cock, drawn from the bottom of the copper, where the coldest water always is, the remaining part, having a greater proportion of fire than before, again begins to boil, though not affected by any increase of heat.

31 See page 267.

32 G. C. stands for great copper, L. C. stands for little copper.

33 Deduction from the first mash for heat created by effervescence and hard corns. See the calculation above.

34 Additions to the mashes on account of heat lost, by the liquor coming from little copper, and by mashing and standing. See page 293.

35 The charge of the first liquor is for 11 barrels, with a deduction of 2 inches, according to the gauges of the coppers, page 221. These two inches answer to the 8 degrees of heat for the effervescence, hard corns, and new malt. See computation above.

36 The second and following mashes are to be charged with as many more inches of boiling water, as answer to the fourth part of the number of degrees of heat lost by the refrigeration of the mashes. See page 294.

37 In beers intended for long keeping, the fermentation is to be governed by the heat of the worts or musts, more than by that of the exterior air.

38 A must or wort, when under fermentation, from its internal motion, increases in heat 10 degrees, and no keeping beers, when under this act, should exceed a heat of 60 degrees; for this reason, worts of this sort should at first be set to ferment at a heat of 50 degrees, and 50 degrees is nearly the mean of the heats these liquors are impressed with, when deposited in cellars, from the time of their being formed, to that of their coming into use. Their long continuance in this state is the reason why six pints of yeast per quarter of malt is a sufficient quantity to be used when the heat of the air is at or below 50 degrees. If, through necessity, processes of this sort are to be carried on when the mean heat of the natural day is more than this, the quantities indicated in the table will be the fittest rule.

39 Though the air bubbles produced from malt liquors are more uniform, as to their size or consistence, than those of natural wines, yet they are not perfectly so; for this reason, and because it requires a greater power to cause

a wort or must of malt to ferment, than it does to keep this act continued, after it is once begun, it is necessary, at first, to apply such a sufficient quantity of yeast as will obtain this purpose; therefore, one half of the remaining six gallons of yeast is put to the wort on its first coming down.

40 The yeast or air bubbles produced from natural wines, vary not only in their consistence, but also in their volume; so that, in their act of fermentation, a progressive effect is the consequence of this want of uniformity. The yeast or air bubbles of barley wines are more uniform; to imitate nature, it is necessary to apply this principle of fermentation by degrees, to cause a progressive effect only. Feeding of drink is the only means to gain this end; thereby the newly applied yeast maintains the drink in its required agitation, in a similar manner as the increased heat and action raised by fermentation causes the air bubbles in natural wines to act and explode, in proportion to their consistence, and to the quantity of elastic air the bubbles contain; and so requisite it is periodically to apply more yeast to this sort of liquor, or regularly to feed it with this enlivening principle, that, in very hot weather, when this, through carelessness, has been omitted, I have known this ale to become foxed or putrefied, and could attribute this accident to no other cause but to a neglect of this sort, as the worts had been regularly brewed, laid thin in the coolers, received all the cold the night could give them, and the tun in which the drink was worked was perfectly clean.

41 I confess this chapter is rather a matter of curiosity, an effusion of fancy, than of any use to me known; if I have suffered it to remain, it has been to shew that when we have long reflected upon a subject, our ideas often lead us beyond power of practice; and with this farther view, that, perhaps, it may become of service in the hands of some more ingenious and more penetrating artist than myself. However, if I trouble my reader with it, it may be said to be in imitation of an author far superior to myself in rank and knowledge.

42 It must be observed, the wines of France in general make the best brandies, and of these, such which justly are termed green wines, (and soon would become acid) this leads us to the nature of the grain, and of the extractions to procure an equal, pure, nutty spirit. Barley, dried scarcely to the denomination of malt, and extracted with the lowest medium, or perhaps one inferior to this, most likely would answer this purpose. I have tried the experiment in a very imperfect manner, and found it answer beyond expectation.

Buy Books Online from
www.Booksophile.com

Explore our collection of books written in various languages and uncommon topics from different parts of the world, including history, art and culture, poems, autobiography and bibliographies, cooking, action & adventure, world war, fiction, science, and law.

Add to your bookshelf or gift to another lover of books - first editions of some of the most celebrated books ever published. From classic literature to bestsellers, you will find many first editions that were presumed to be out-of-print.

Free shipping globally for orders worth US$ 100.00.

Use code "Shop_10" to avail additional 10% on first order.

Visit today
www.booksophile.com

Milton Keynes UK
Ingram Content Group UK Ltd.
UKHW040656191223
434651UK00003B/326